T0342430

Real-Time Ground-Based Flight Data and Cockpit Voice Recorder

Real-Time Ground-Based Flight Data and Cockpit Voice Recorder

Implementation Scenarios and Feasibility Analysis

Mustafa M. Matalgah
Department of Electrical and Computer Engineering
University of Mississippi
Oxford, MS, USA

Mohammed Ali Alqodah
Department of Electrical and Computer Engineering
University of Mississippi
Oxford, MS, USA

IEEE PRESS
WILEY

Published by John Wiley & Sons, Inc., Hoboken, New Jersey.
Published simultaneously in Canada.

For general information on our other products and services or for technical support, please contact our Customer Care Department within the United States at (800) 762-2974, outside the United States at (317) 572-3993 or fax (317) 572-4002.

Wiley also publishes its books in a variety of electronic formats. Some content that appears in print may not be available in electronic formats. For more information about Wiley products, visit our web site at www.wiley.com.

Library of Congress Cataloging-in-Publication Data

Names: Matalgah, Mustafa M., author. | Alqodah, Mohammed Ali, author.
Title: Real-time ground-based flight data and cockpit voice recorder : implementation scenarios and feasibility analysis / Mustafa M. Matalgah, Mohammed Ali Alqodah.
Description: Hoboken, New Jersey : Wiley, [2024] | Includes bibliographical references and index.
Identifiers: LCCN 2023043757 (print) | LCCN 2023043758 (ebook) | ISBN 9781119984863 (hardback) | ISBN 9781119984870 (adobe pdf) | ISBN 9781119984887 (epub)
Subjects: LCSH: Flight recorders. | Cockpit voice recorders. | Aircraft accidents.
Classification: LCC TL589.2.F5 M38 2024 (print) | LCC TL589.2.F5 (ebook) | DDC 629.135–dc23/eng/20231102
LC record available at https://lccn.loc.gov/2023043757
LC ebook record available at https://lccn.loc.gov/2023043758

Cover Design: Wiley
Cover Image: © bluebay/Shutterstock

Set in 9.5/12.5pt STIXTwoText by Straive, Chennai, India

This book is dedicated to our beloved mothers and the cherished memory of our fathers, as well as to our dear families. Your profound impact on our lives is immeasurable, and we are forever grateful for the remarkable gift of family.

Mustafa M. Matalgah
Mohammed Ali Alqodah

Contents

About the Authors

Mustafa M. Matalgah obtained his undergraduate and master's education from Jordan and his PhD from the University of Missouri, Columbia, United States, all in electrical engineering. He has a wide range of academic and industry experiences in the electrical engineering field with emphasis on communication engineering. From 1996 to 2002, he was with Sprint Corporation, Kansas City, MO, United States, where he held various technical positions leading a wide range of projects dealing with optical communication systems deployment and the evaluation and assessment of emerging wireless communication technologies. Since August 2002, he has been with the University of Mississippi where he is now a full professor of electrical engineering. In Summer 2008, he was a visiting professor at Chonbuk National University in South Korea. He was also a visiting professor and program evaluator at Misr International University (MIU) in Egypt in Summers 2009, 2010, and 2012. He also held academic positions in Saudi Arabia and short-term positions in Jordan. His current technical and research experience is in the performance evaluation and optimization of wireless communication systems in emerging technologies. He has more than 150 archival publications (including journals, conference proceedings, book chapters, and patents) in addition to more than two dozens of industry technical reports in these areas. He served as the research supervisor of, and served on defense committees on, several MSc and PhD students. He received several certificates of recognition for his work accomplishments in industry and academia. He is the recipient/co-recipient of the Best Paper Award on several international and regional conferences and workshops. He is the recipient of the 2006 School of Engineering Junior Faculty Research Award at the University of Mississippi. He served on the Editorial Board of a few international journals, served as member and chair on several university committees, chair on several international conferences sessions and workshops, member on several international conferences technical program and organizing committees, reviewer for several funding proposals in United States and Canada, and project manager on several projects in the industry. He served on the Faculty Senate of the University of Mississippi for four years.

Mohammed Ali Alqodah holds a master's degree in electrical engineering/ communications and electronics engineering from the Jordan University for Science and Technology, Irbid, Jordan, which he attained in March 2008. Prior to that, he earned his bachelor's degree in the same discipline from the same university in June 2004. His academic journey is currently culminating with his PhD program at the University of Mississippi, where he is continually advancing his research skills and contributing to cutting-edge developments in the field of electrical and computer engineering. His research interests span diverse areas, including wireless communication, image processing, nonlinear theory, and optical communication. He has gained significant experience in the academic field, serving as a teaching assistant at the University of Mississippi since August 2021. Additionally, he was a Lecturer at the Electrical Engineering Department of Prince Sattam bin Abdulaziz University in KSA from September 2009 to June 2021. During his tenure, he conducted various courses and labs, imparting knowledge in Signal and Systems, Communications Systems, Digital Signal Processing, Wireless Communications, Electromagnetics, Microwave Engineering, and more. His passion for education is further evident through his active participation in various committees. As an accomplished researcher, he has contributed significantly to his field, with several of his publications being recognized for their impact. In addition to his academic pursuits, he actively engages in professional service, serving as a Technical Reviewing Committee member for esteemed international conferences and journals.

Foreword

This book presents information on flight data and cockpit voice recorders that are fundamental to the investigations of aircraft accidents and flight abnormality incidents. Starting with a motivation explaining the inadequacy of usefulness of hardware recorders onboard the aircraft, in case of long delays in locating them or missing/damaged recorders after aircraft crashes overseas or inaccessible terrains, the authors proceed to deliver in next seven chapters details of the current recording systems, current aviation communications systems technology, utility of satellite, unmanned aerial vehicle (UAV) and other cooperative communications for transmitting in-flight data in real-time to a ground-based server facility, and future of aviation communications in gathering real-time data safely and securely. Professor Matalgah, with his expertise in wireless communications and in collaboration with his graduate student Mr. Alqodah, is able to render the book concisely within 150 pages. This book should be useful to general readers who have an interest in science and technology as well as to engineers and technicians working in aviation electronics and communications technology.

Dr. Ramanarayanan Viswanathan
Professor and Chair
Department of Electrical and Computer Engineering
University of Mississippi
Oxford, MS, USA

In the rapidly evolving sphere of aviation, a domain where safety is paramount, the quest for relentless innovation and progressive development is a constant. Nestled within this incessant journey toward enhanced safety emerges a groundbreaking piece – *Real-Time Ground-Based Flight Data and Cockpit Voice Recorder: Implementation Scenarios and Feasibility Analysis*. This is not just a book but a revolutionary paradigm that stands distinguished, illuminating uncharted territories in aviation safety.

Written with intricate detail and profound expertise, this seminal work serves not just as a manuscript but as an essential compass for aviation specialists, academics, and aficionados alike. It is an invitation to delve into the depths of the transformative impact that real-time data transmission harbors for augmenting aviation safety standards globally.

Authored by the esteemed Professor Matalgah and Mr. Alqodah, this book transcends the conventional narratives and ideologies. It is an assemblage of disruptive ideas that challenge status quo thinking, unveiling an enlightened pathway toward a safety ecosystem that is empowered and enhanced by the immediacy and accessibility of real-time data.

My exploratory journey through the vast expanse of existing literature and research in this domain bore no resemblance to the innovative discourse presented in this book. The authors have managed to encapsulate complex challenges and ingenious solutions with an elegance and thoroughness that is unparalleled. The discourse is not just compelling but emerges as a pressing narrative in the contemporary context, demanding the attention of professionals and experts spanning airplane safety systems, airline safety protocols, and aero communication research labs.

Real-Time Ground-Based Flight Data and Cockpit Voice Recorder: Implementation Scenarios and Feasibility Analysis is a pioneering endeavor. In a world where information on this pivotal topic is dispersed and fragmented, this book encapsulates, organizes, and presents insights with a clarity and coherence that is distinctly absent in existing literature. The implications of this work are far-reaching; it stands poised to ignite dialogs, foster collaborations, and propel advancements that could redefine the contours of aviation safety.

In essence, this is more than a commendable read – it is a monumental contribution that could potentially recalibrate our approach to, and expectations of, safety in the multifaceted world of aviation. Every page is imbued with the promise of a future where aviation safety is not just an organizational commitment but a global, collaborative, and innovative venture that harnesses the potency of real-time data in ways previously unimagined.

Dr. Farid Nait-Abdesselam
Professor of Computer Science
University of Missouri Kansas City
Kansas City, MO, USA

This book serves as a comprehensive guide to the evolving landscape of aviation safety through real-time ground-based flight data and cockpit voice recorder solutions. By examining existing challenges, exploring innovative technologies, and envisioning future advancements, it paves the way for a safer and more effective aviation industry. This book marks a pioneering endeavor, introducing innovative concepts and solutions that are poised to capture the attention and interest of various stakeholders within the aviation industry.

In the aviation industry where safety is paramount, this work aims to address existing challenges in traditional recorders and explore innovative approaches to enhance airplane crucial data transmission. This book embarks on a journey to introduce an alternative paradigm – real-time ground-based flight data recorder and cockpit voice recorder (FDR/CVR) implementation. By shifting the focus from on-board recorders to ground-based solutions, the aim is to overcome the limitations of traditional methods. The book also ventures into the future of aviation communication, unveiling advancements that hold promise for improving the streaming of FDR and CVR data. The introduction of developments in air-to-ground and air-to-air communication and the integration of machine learning are all explored as potential pathways to reshape aviation communication for heightened safety and efficiency.

I recommend this book for its groundbreaking insights into aviation safety, featuring real-time integration of flight data and cockpit voice recorders and setting a transformative course for the industry.

Dr. Majdi Bsoul
Data Science and Analytics Practice Lead
Nokia
Dallas, TX, USA

Preface

The skies have always captivated humanity's imagination, and aviation has become an integral part of our modern lives. Countless hours of meticulous effort have been invested in enhancing the safety standards of aircraft, whether for civilian, commercial, or military purposes. For aviation history, safety measures have evolved significantly, transforming air travel into one of the safest modes of transportation today. Despite this remarkable progress, the aviation industry remains committed to comprehending the intricacies of aviation accidents and finding ways to prevent them from occurring in the first place. The study and analysis of aircraft incidents and crashes are pivotal to achieving this objective.

Central to the endeavor of investigating aircraft accidents is the flight data and cockpit voice Recorder (FDR/CVR), often referred to as the "Black Box." This electronic marvel, housed within the aircraft, records critical information about the aircraft's operational parameters and cockpit conversations. By scrutinizing the data stored within the FDR/CVR, investigators endeavor to unravel the chain of events leading up to an accident, discerning the root causes that contributed to the tragedy. This wealth of data plays an indispensable role in understanding and addressing aircraft malfunctions and accidents.

However, the path to uncovering these invaluable insights has not always been smooth. There have been instances where the retrieval of FDR/CVR from crash sites posed significant challenges, leaving investigators without this crucial tool for deciphering accident causes. Throughout the years, numerous aviation accidents have involved difficulties in recovering FDR/CVRs, impeding the thorough analysis needed for comprehensive accident investigations.

As technology advances, so does the potential to refine aviation safety practices. This book explores the domain of real-time ground-based FDR/CVR, offering an innovative approach to augment existing FDR/CVR technology. By exploring alternative methodologies for data transmission, we aim to address the limitations faced by traditional on-board recorders, especially in cases where access to crash sites is restricted or delayed.

In "Real-Time Ground-Based Flight Data and Cockpit Voice Recorder: Implementation Scenarios and Feasibility Analysis," we embark on a journey through the intricate landscape of aviation safety. This book investigates the challenges and prospects of implementing real-time data streaming from aircraft to ground stations. We examine existing technologies, explore cutting-edge communication systems, and analyze the feasibility of deploying innovative solutions to enhance aviation safety.

Chapter 1: The introduction highlights the crucial significance of aviation safety and the role of FDR/CVRs in accident investigations. We investigate the history of safety enhancements and the ongoing necessity to understand accidents comprehensively, aiming to prevent potential disasters ahead.

Chapter 2: State of the Art takes a comprehensive look at the current landscape of FDR/CVR technology and its challenges. We examine the potential of real-time data streaming and review existing developments in flight data transmission systems.

Chapter 3: Aviation Communication Overview offers insights into the evolution of aviation communication systems from their inception to the present day. We explore the diverse forms of communication in the aviation realm and the pivotal role of various stakeholders.

Chapter 4: The potential of satellite-based data transfer for FDR/CVRs is explored. We assess the capabilities of different satellite systems and their applicability in ensuring real-time data transfer.

Chapter 5: VHF Digital Link Implementation introduces the concept of very high-frequency digital Link (VDL) and its potential to transmit vital flight data. We examine the viability of VDL implementation and its technical details.

Chapter 6: Cooperative Data Transmission Implementations explores the concept of cooperative communication among aircraft to overcome limitations in direct communication methods. We delve into the potential of collaborative data transmission to enhance aviation safety.

Chapter 7: Unmanned Aerial Vehicles (UAVs) play a key part in wireless communication networks, as explored in the UAV Wireless Networks and Recorders chapter. We discuss the deployment of real-time recorders on airplanes using UAVs and their significance in accident investigations.

Chapter 8: Future Aviation Communication provides a glimpse into the promising future of aviation communication. We explore advancements such as System Wide Information Management (SWIM), cutting-edge air-to-ground and air-to-air communication technologies, and the integration of machine learning for enhanced communication.

This book is a journey of discovery, innovation, and determination to make aviation even safer. It explores of how technology can reshape the way we understand and prevent accidents, offering a fresh perspective on the vital interplay between data, communication, and aviation safety. Through each chapter, we explore deeper into the challenges and opportunities that lie ahead to make the skies safer for future generations.

Department of Electrical and Computer
Engineering
University of Mississippi
Oxford, MS, USA

14 August 2023

Mustafa M. Matalgah
Mohammed Ali Alqodah

Acknowledgments

We express our heartfelt gratitude to Situmbeko A. Matale for his invaluable contributions to this book. His diverse expertise and efforts were instrumental in shaping Chapters 4–6. Situmbeko played a pivotal role in providing in-depth discussions, conducting numerical validations, and creating essential figures for the first-generation Iridium satellite data transfer implementation, the VDL Mode 4 implementation, the VHF and satellite system cooperation, and VDL system-based relaying. His valuable input has greatly enriched this work, and we sincerely appreciate his dedication and support.

Mustafa M. Matalgah
Mohammed Ali Alqodah
Department of Electrical and
Computer Engineering
University of Mississippi
Oxford, MS, USA

Acronyms

AAC	airline administration communications
AANET	aeronautical ad-hoc network
ACARS	Aircraft Communications Addressing and Reporting System
ACAS	airborne collision avoidance system
ACKs	send Acknowledgments
ADS-B	automatic dependent surveillance-broadcast
ADS-C	aircraft position reporting for air traffic control
ADS-R	automatic dependent surveillance – rebroadcast
AeroMACS	aeronautical mobile communication system
AFN	air traffic services facilities notification
AI	artificial intelligence
ALOHA	additive links on-line Hawaii area
AM	amplitude modulation
AMBE	advanced multi-band excitation
AMSK	amplitude modulated minimum shift keying
ANS	air navigation system
APD	Aireon Processing and Distribution
ARINC	Aeronautical Radio Incorporated
ASAS	airborne separation assurance systems
ATC	air traffic control
ATCRBS	air traffic control radar beacon system
ATSCs	air traffic service communications
CPDLC	controller pilot data link communications
CR	cooperative radiation
CRC	cyclic redundancy check
CRN	cognitive radio networks
CSMA	carrier sense multiple access
CTAF	common traffic advisory frequency

CVR	cockpit voice recorder
DDM	difference in depth modulation
DEQPSK	differentially encoded QPSK
DLS	data link service
DSPs	datalink service providers
EGC	enhanced group call
ELM	extended-length message
EUROCAE	European Organisation for Civil Aviation Equipment
FAA	Federal Aviation Administration
FANET	flying ad-hoc network
FANS	future air navigation system
FDAMS	flight aata acquisition and management systems
FDAU	flight data acquisition unit
FDM	frequency division multiplexing
FDR	flight data recorder
FDS	flight deck safety
FDX	full duplex
FEC	forward error correction
FSO	free-space optical
GBAS	ground based augmentation system
GFSK	Gaussian frequency shift keying
GPS	global positioning system
HF	high frequency
HFDL	HF data link
IBN	infrastructure-based network
ICAO	International Civil Aviation Organization
ILN	infrastructure-less network
ILS	instrument landing system
IMP	information management panel
IoT	Internet of Things
LAAS	local area augmentation system
LEO	low earth orbit
LME	link management entity
LoRa	long range
LOS	line of sight
LUF	lowest usable frequency
MAC	media access control
MANET	mobile ad-hoc networks
MEC	mobile edge computing
METAR	meteorological aerodrome report
MILP	Mixed Integer Linear Programming

MIMO	multiple-input multiple-output
mmWave	millimeter wave
M-PSK	M-ary phase shift keying
MSC	mobile switching center
MUF	maximum usable frequency
NAS	national airspace system
NOMA	non-orthogonal multiple access
NTSB	National Transportation Board
NTSC	National Transportation Safety Committee
NVIS	near visual interface
OOOI	out, off, on, in format
POA	plain old ACARS
PPM	pulse position modulation
PSTN	public switched telephone network
QPSK	quadrature phase shift keying
RAs	resolution advisories
REQ	resend requests
RHC	right hand circular
RTCA	radio technical commission for aeronautics
SARPS	standards and recommended practices
SATCOM	satellite communications
SB-S	SwiftBroadband-Safety
SDN	software-defined networking
SD-WFR	software-defined wireless flight recorder
SITA	international company for aeronautical telecommunications
SNAcP	subnetwork access protocol
SNR	signal-to-noise ratio
SPOs	single-pilot operations
SSB	single sideband
STDMA	self-organized TDMA
SWIM	system wide information management
TCAS	traffic alert and collision avoidance system
TDD	time division duplex
TDM	time division multiplexing
TIS-B	traffic information service – broadcast
TT&C	telemetry, tracking, and commanding
UAs	unmanned aerial vehicle
UAT	universal access transceiver
UAVCN	unmanned aerial vehicle communication network
ULB	Underwater Locator Beacon
VANET	vehicular ad-hoc networks

VDB	VHF data broadcast
VDL	very high-frequency digital link
VHF	very high frequency
VOR	VHF omnidirectional radio range
WMNs	wireless mesh networks
WSN	wireless sensor network

1

Introduction

Aviation safety in air travel has always been a vital part of the aviation industry. Endless man-hours have gone into improving the safety standards of aircraft for civilian, commercial, and military aviation. Since the inception of commercial aviation, industry safety standards have improved dramatically, making flying one of the safest modes of transportation today. Even with a great record of safety today, measures are still undertaken to ensure that aviation accidents are not only well understood but also that they may, in the future, be made avoidable. Studying and understanding the causes of aircraft incidents and crashes is one of the main methods used to ensure that transportation in the sky is as safe as possible. One of the main methods used by investigators and engineers to study the causes of aircraft accidents is the use of the flight data and cockpit voice recorder (FDR/CVR), also known as the Black Box. However, FDR/CVR has not been efficiently useful in some catastrophic accidents such as the recent Aeroflot Flight 1492 (a Sukhoi Superjet 100) that was operating a domestic flight in Russia [Hradecky, 2019], the Lion Air Flight 610 (a Boeing 737 MAX 8) that crashed into the Java Sea shortly after takeoff from Soekarno-Hatta International Airport in Jakarta [National Transportation Safety, 2018], the EgyptAir Flight 804 (an Airbus A320) that crashed into the Mediterranean Sea [Yeung, 2016], and the missing Malaysia Airlines Flight MH370 (a Boeing 777-200ER) that disappeared on 8 March 2014 [MacLeod et al., 2014], just to name a few. Throughout this book, we will explore a real-time ground-based FDR/CVR as an alternative to the existing FDR/CVR.

1.1 Motivation

The FDR/CVR is an electronic device that is installed on board an aircraft and is used to record important details about the status and performance parameters of

Real-Time Ground-Based Flight Data and Cockpit Voice Recorder: Implementation Scenarios and Feasibility Analysis, First Edition. Mustafa M. Matalgah and Mohammed Ali Alqodah.
© 2024 The Institute of Electrical and Electronics Engineers, Inc. Published 2024 by John Wiley & Sons, Inc.

the aircraft. Accident investigators use the data stored in the FDR and CVR to try and piece together the events preceding the accident to pinpoint what went wrong with the flight. The information contained in the FDR and CVR plays an extremely important role in the investigation of airplane malfunctions and crashes. However, there have been incidents in the past where investigators were unable to recover FDR and CVRs from crash sites. This leaves them without one of their most valuable resources in discovering the causes of a catastrophe. Reports have shown that since 1980, there have been nearly 30 aviation accidents in which the flight recorder was either not found or damaged, and hence data was not recovered. On the other hand, in some commercial aviation accidents over water, the recovery time delay of the FDR/CVR could range from a week to a couple of years. This adds heavy costs to the search process and brings danger to rescuers lives in some situations, like what happened during the Lion Air Flight 610 (MAX 8), when one member of the volunteer rescue team died during the CVR recovery operations. In the following, we lay out detailed examples of catastrophic flight accidents that involved such kinds of drawbacks in the recovery work.

On 5 May 2019, Aeroflot Flight 1492 [Hradecky, 2019], a Sukhoi Superjet 100 operating a domestic flight in Russia, suffered an inflight electrical failure shortly after departing from Sheremetyevo International Airport, Moscow, and returned to the airport where it caught fire after landing; 41 of the 78 people on board died. The CVR was found in satisfactory condition, but the FDR casing was damaged by exposure to extremely high temperature. On October 29, 2018, Lion Air Flight 610 [National Transportation Safety, 2018], a Boeing 737 MAX 8, crashed into the Java Sea shortly after takeoff from Soekarno-Hatta International Airport in Jakarta en route to Depati Amir Airport in Pangkal Pinang, Indonesia. All 181 passengers and eight crew members were killed. After two days, on 1 November 2018, of searching operations, the flight's FDR was recovered, which was located at a depth of 32 m (105 ft) underwater, while the CVR was reported as not found. During these two days of recovery operations, one member of the volunteer rescue team died. Lion Air paid US$2.8 million for a second attempt to search for the CVR between 19 and 29 December. The Indonesian National Transportation Safety Committee (NTSC) funded a further underwater search operation using the Indonesian Navy vessel KRI (kapal perang Republik Indonesia) Spica, which started on 8 January 2019 and continued until the CVR was recovered on 14 January. The CVR was found at a depth of 30 m (98 ft) covered by mud that was 8 m (26 ft) thick. The third example of these kinds of catastrophes is EgyptAir Flight 804 (Airbus A320) [Yeung, 2016], a regularly scheduled international passenger flight from Paris Charles de Gaulle Airport to Cairo International Airport operated by EgyptAir, which on 19 May 2016 crashed into the Mediterranean Sea, killing all 56 passengers, 3 security personnel, and 7 crew members on board. After a multinational search and recovery operation, the flight recorders were recovered after 10 days (29 June 2016) resting on the sea floor, 3000 meters down. Another example is the most notable deadliest incident involving Boing 777 and the deadliest in Malaysia airline history is the Malaysia Airlines Flight (MH370) [MacLeod et al., 2014] that disappeared on 8 March 2014, and all 239 passengers and crew were presumably missing. The CVR and FDR recorders have not been found until today (as we are writing this book). On 6 October 2014, the governments of Malaysia, China, and Australia made a joint commitment to search 46,000 mi^2 of the seafloor thoroughly, but after nearly three years of combing the far desolate Indian Ocean and its deep seabed of a 46,000 mi^2 zone without finding the missing Boeing 777. Then the governments of Malaysia, Australia, and China called off the most complex and expensive search in aviation history. In June 2014, Time (https://time.com/2854385/mh370-search-spending/) estimated that the total search effort up to that point had cost nearly US$70 million. Malaysia's Ministry of Transport revealed that it had spent 280.5 million Malaysian ringgit (US$70 million) on the research as of February 2016. Some search plans were proposed to the Malaysian government during the period 2017–2022. Here it is worthwhile to quote the statement: "Despite every effort using the best science available, cutting-edge technology, as well as modeling and advice from highly skilled professionals who are the best in their field, unfortunately, the search has not been able to locate the aircraft", the Joint Agency Coordination Center in Australia said in a statement [Denyer, 2017]. On the contrary, Ocean Infinity, with the Malaysian government's approval, declared in January 2018 that it would resume the search in a narrow 25,000 km^2 area [Associated Press, 2018]; however, on 9 June 2018, it was revealed that the search had been unsuccessful [Gartland, 2019]. In March 2019, Ocean Infinity stated again that it was ready to resume the search. They believed that the most probable location was still somewhere along the seventh arc around the area identified previously and upon which its 2018 search was based. In March 2022, they committed to resuming their search in 2023 or 2024, pending approval by the Malaysian government.

As seen from the previous examples, it is obvious that traditional CVR/FDR recorders are not effectively reliable; they are either not found, damaged, or recovered late. Even in the cases when the FDR/CVR recorders are found, the recovery operations were reported to be costly and took a long time. Table 1.1 summarizes commercial aviation accidents over water between the years 2000 and 2015, for which the FDR or CVR recovery time was a week or more. As seen in the table, in some accidents, it took around two years to find the FDR/CVR recorders.

Flight-recorded data could be used further if it were made available to technicians and specialists on the ground while the plane is still in-flight to possibly prevent catastrophe. In addition to being outdated, storing flight data onboard the aircraft is ineffective and limited by the quantity of data stored. Today's wireless telecommunication technologies can be utilized to achieve the goal of

Table 1.1 Commercial aviation accidents over water from 2000 to 2015 where the recovery of the CVR or the FDR took a week or more.

Accident date	Flight Name/ number	Aircraft type	Accident location	Phase	FDR recovery time	CVR recovery time
28 December 2014	AirAsia Indonesia QZ8501	Airbus A320	Karimata Strait, Java Sea	En-Route	16 days	17 days
25 January 2010	Ethiopian Airlines Flight 409	Boeing B737	Mediterranean Sea	Take off	14 days	22 days
30 June 2009	Yemenia Flight 626	Airbus A310	Moroni, Comoros Islands	Approach Approach	60 days	60 days
1 June 2009	Air France Flight 447	Airbus A330	Atlantic Ocean	en route	1 year 11 months and 2 days	1 year 11 months and 3 days
15 January 2009	US Airways Flight 1549	Airbus A320	New York, United States	Climb	7 days	7 days
9 August 2007	Air Moorea Flight 1121	DHC6	Off the coast of Moorea	Approach	N/A[a] N/A[a]	21 days 21 days
1 January 2007	Adam Air Flight 574	Boeing B737	Makassar Strait	en route	240 days	240 days
2 May 2006	Armavia Air Flight 967	Airbus A320	Black Sea	Approach	22 days	20 days
6 August 2005	Tuninter Flight 1153	ATR-72	Off the coast of Palermo, Italy		24 days	23 days
25 June 2002	China Airlines Flight 611	Boeing B747	Taiwan Strait	Climb	25 days	24 days
7 June 2002	China Northern Airlines Flight 6163	MD-82	Bay near Dalian, China		14 days	7 days
30 January 2000	Kenya Airways Flight 431	Airbus A310	Abidjan	Take off	6 days	26 days

a) Aircraft is too small and not carrying FDR recorders.
Source: Adapted from NTSB [n.d.].

transmitting real-time (up to the minute) flight information data from aircraft to stations on the ground, creating a ground-based mirror of FDR/CVR recorders. These ground-based stations can be used by technicians not only for data storage but also for data analysis and monitoring as well as technical support.

In this book, an alternative method of storing this critical aircraft data is explored. Making the instant FDR/CVR data available by transmitting flight data instantaneously to a ground base station has received great interest by researchers and manufacturers and examined by industry specialists as an alternative to the traditional methods during the last two decades, especially after the Malaysian MH370 flight disaster in 2014.

The advantages of the alternative real-time flight recorder concept that has received attention recently are tremendously superior to the existing traditional FDR/CVR flight recorder. A few of the motivations for replacing the FDR and CVR system include:

1. Recovery of FDR and CVR is costly and time-sensitive.
2. If FDR and CVR cannot be retrieved, the investigation into the crash can be greatly hampered. Information stored at a remote location does not have to be limited to 30 minutes and 25 hours for CVR and FDR, respectively, potentially providing more feasible insight into the events that lead to catastrophes.
3. Transmitting real-time aircraft data (streaming), during and momentarily before aircraft emergencies, can help ground-based personnel assist pilots and crew more effectively.
4. Transmitting real-time aircraft state data can be monitored by an automated system at the ground station, alerting the flight crew of events that may not be routinely monitored by them.
5. Offsite storage can allow data to be stored indefinitely and used by maintenance crews on the ground during aircraft service.

1.2 Entities Involved in Air Crash Investigations

Various entities play a role in air crash accident investigations. This section provides an overview of the key players involved in air crash accident investigations in the United States and explains their respective roles and responsibilities in the process.

1.2.1 Federal Aviation Administration (FAA)

The FAA plays a critical role in ensuring the safety of air travel in the United States. As a regulatory agency operating under the Department of Transportation (DOT), it has the authority to oversee all aviation-related activities, including air traffic control, pilot, and crew certification, and aircraft manufacturing and maintenance. The FAA's involvement in air crash investigations is primarily focused on specifying the necessary equipment requirements for FDR and ensuring proper handling of the recorded data, with particular responsibility to outline the equipment requirements for these recorders, including the type of data they should capture, how the data should be stored and retrieved, and how the data should be protected from damage or tampering. On the other hand, the FAA faces the challenge of balancing its responsibility for promoting aviation and air commerce with its priority of ensuring flight safety. These two goals can sometimes create conflicting interests. Despite these challenges, the FAA remains committed to its primary objective of ensuring the safety of air travel in the United States.

1.2.2   National Transportation Board (NTSB)

The NTSB is an independent federal agency with the mission and authority to investigate all civil aviation accidents in the United States, as well as significant accidents in other modes of transportation, such as railroads, highways, marine vessels, and pipelines. The NTSB issues safety recommendations to prevent future accidents. In air crash investigations, the NTSB plays a crucial organizational role. Its responsibility is to conduct the crash investigation, determine the probable cause of the accident, and make recommendations to enhance aviation safety. The FAA receives the NTSB's recommendations but is not legally bound to act upon them. To carry out its investigation, the NTSB deploys field investigation teams to the crash site to collect all available information. The NTSB also maintains laboratories that analyze crash data collected from crash sites, including flight recorder information. One of the key tasks of the NTSB is to determine the "probable cause" of the accident by analyzing multiple factors, such as weather, flight crew actions, training, medical conditions, aircraft maintenance, and air traffic control. By examining these factors, the NTSB is almost always able to determine the sequence of events in the accident chain and determine the probable cause. The availability of flight data is critical to accident investigation. Flight data is usually retrieved from flight recorders on board the accident aircraft or from a database on the ground where real-time flight data was transmitted to. The data must be accurate, complete, and uncompromised, and it must be available in a timely manner. Any system for real-time transmission of flight data must deliver timely, accurate, and complete data to the NTSB.

1.2.3   Operator (Airline)

The role of the airline operator in air crash investigations is to collaborate with both the FAA and the NTSB by providing the flight recorder data from the accident flight. As they are directly involved in the aviation industry and responsible for ensuring the safety of their passengers and crew, the operator has a vested interest in the investigation and will provide any necessary support and assistance to determine the probable cause of the accident. They will also work toward modifying procedures and practices in the interest of enhancing flight safety. Since operators manage numerous flights daily, their flight recorders collect a significant volume of flight data, which is considered their property until it is released in accordance with NTSB 830 and 14 Code of Federal Regulations (CFR) regulations. The operator's responsibility is to ensure that the data is accurate and complete before it is released to the NTSB. The timely provision of this data is crucial to the investigation process, and the operator must cooperate with the NTSB to facilitate the release of the data.

1.2.4   Equipment Manufacturer

Equipment manufacturers, including prominent companies like Boeing, Airbus, and Embraer, have a strong vested interest in building safe and reliable aircraft. In the context of accident investigation, their role is to provide technical expertise and support as needed. This may involve offering details about the accident aircraft or conducting engineering tests on recovered components. They may also participate in technical discussions in an advisory capacity. During the design and manufacturing of airframes, equipment manufacturers play a crucial role in ensuring that flight data sensors are properly placed, that there is a pathway for the sensors to communicate with the flight recorders, and that space is provided for the installation of flight recorders. However, it is important to note that aircraft manufacturers themselves do not actually make flight recorders. Instead, these devices are purchased from separate vendors and installed in the airframe by the manufacturer or its customer. While aircraft manufacturers are responsible for installing the flight recorders required by the FAA, they do not dictate the nature of the recorders or the data that they must collect.

1.3   Existing Traditional FDR/CVR

The FDR and the CVR have been vital tools in air crash investigations for decades. The FDR records crucial flight data parameters, including altitude, airspeed, heading, vertical acceleration, and more. The CVR, on the other hand, captures cockpit audio, including conversations between the pilots and air traffic control, and other sounds such as warning alarms. Together, the FDR and CVR provide a comprehensive record of what happened in the cockpit leading up to an incident or accident. These devices have evolved significantly over time, from early models that used photographic film and metal foil to the latest solid-state recorders with high recording capacity and durability. In this section, we will provide a detailed description of both the FDR and the CVR, including their functions, features, and capabilities.

1.3.1 Traditional FDR/CVR History

The concept of aircraft FDR existed long before human-powered flight. In the early days, basic data was described or recorded by handwritten notes on altitude and wind direction [Fisher, 2017]. In their inventions, the Wright brothers used a crude instrument that measured time, distance, and engine RPMs. Many airplanes in the 1920s had small barographs with a rotating paper drum and ink stylus. In 1939, a method for capturing flight information about an aircraft, such as altitude, air speed, and the location of the cockpit controls, was created by François Hussenot [Engber, 2014]. Through calibrated mirrors, onboard sensor readings are flashed into a box of photographic film and traced to a running tab of flight parameters. In the mid-1950s, two models of flight recorders were constructed, the General Mills Ryan Flight Recorder and the Australian Research Laboratories Flight Memory Unit [Dub and Parizek, 2018]. General Mills (GM) Flight Recorder was based on a patent by a mechanical engineering professor (James J. Ryan) at the University of Minnesota [Dub and Parizek, 2018], and it was capable of storing four data parameters (velocity, g-force, altitude, and time) for up to 300 hours using a needle engraving into the metal foil. The other model was presented on a four-page note submitted in 1954 that opens the first chapter of the history of flight recorders; Dr. David Warren proposed the idea of a device that would record not only flight data but also voices and other cockpit noises just before a crash and save them on a running tab [Marsh, 2021] (Figure 1.1). In his idea, Warren outlined how crucial information may be obtained from the moments just before an accident and made available to crash investigators. He suggested a device for recording cockpit conversations onto a closed circle of wire. He advised placing the recorders in an area of the plane that is least likely to sustain significant damage, particularly the tail, to aid recovery after a crash.

After many fatal crashes in the mid-1950s that left no survivors or witnesses, implementing FDRs into big commercial aircraft became vital. However, it was not until the mid-1960s that FDR/CVR recorders were mandated as a requirement in the commercial aircraft industry. The second generation of flight recorders, which were in use during the 1970s and 1980s, relied on the idea of digital magnetic recording onto a wire of a plastic or metallic magnetic strip. The most important development in flight recorder technology came with the introduction of solid-state recorders in the late 1980s. Recorders' dependability, crash and fire resistance, and recording capacity have all increased thanks to the adoption of solid-state memory in flight recorders. There are currently two-hour audio CVRs

Figure 1.1 Dr. David Warren with the first prototype flight data recorder, 2004.
Source: National Museum Australia/Wikimedia Commons/Public Domain.

and digital FDRs that can record up to 256 12-bit data words per second, four times as fast as digital FDRs that use magnetic tape [SKYbrary, 2014].

It is worth noting that the FDR/CVR recorder (black box) is usually painted in bright colors (like orange and red) [Engber, 2014] to improve its visibility if it needs to be recovered amid airplane wreckage. The primary role of the FDR/CVR is to aid investigators in determining the cause of a crash or incident involving an aircraft. They are designed to have high durability and to be preserved, in cases when an incident destroys the aircraft. The FDR and CVR consist of two pieces of equipment. These are:

1. The Flight Data Recorder (FDR)
2. The Cockpit Voice Recorder (CVR)

Together, the FDR and the CVR are also referred to as flight recorders. In the sequel, a detailed description of both the FDR and the CVR is provided.

1.3.2 Flight Data Recorder (FDR)

The FDR is a device that records information about the aircraft's performance and status. Sensors installed in and around the aircraft monitor performance-related aspects of the aircraft and flight and transmit data to the flight data acquisition unit (FDAU), a device that formats the data before sending it to the FDR for storage. These sensors monitor and read data. Currently, FAA and NTSB in the United States require that the system monitor and ultimately record 88 specific parameters [Grossi, 2006]:

1. Timestamp;
2. Pressure altitude;
3. Indicated airspeed;
4. Heading – primary flight crew reference (if selectable, record discrete, true, or magnetic);
5. Normal acceleration (vertical);
6. Pitch attitude;
7. Roll attitude;
8. Manual radio transmitter keying, or CVR/digital flight data recorder (DFDR) synchronization reference;
9. Thrust/power of each engine – primary flight crew reference;
10. Autopilot engagement status;
11. Longitudinal acceleration;
12. Pitch control input;
13. Lateral control input;
14. Rudder pedal input;
15. Primary pitch control surface position;
16. Primary lateral control surface position;
17. Primary yaw control surface position;
18. Lateral acceleration;
19. Pitch trim surface position;
20. Trailing edge flap or cockpit flap control selection;
21. Leading edge flap or cockpit flap control selection;
22. Each Thrust reverser position (or equivalent for propeller airplane);
23. Ground spoiler position or speed brake selection;
24. Outside or total air temperature;
25. Automatic Flight Control System (AFCS) modes and engagement status, including autothrottle;
26. Radio altitude (when an information source is installed);
27. Localizer deviation, Microwave Landing System (MLS) azimuth;
28. Glideslope deviation, MLS elevation;
29. Marker beacon passage;
30. Master warning;
31. Air/ground sensor (primary airplane system reference nose or main gear);
32. Angle of attack (when information source is installed);
33. Hydraulic pressure low (each system);

34. Ground speed (when an information source is installed);
35. Ground proximity warning system;
36. Landing gear position or landing gear cockpit control selection;
37. Drift angle (when an information source is installed);
38. Wind speed and direction (when an information source is installed);
39. Latitude and longitude (when an information source is installed);
40. Stick shaker/pusher (when an information source is installed);
41. Windshear (when an information source is installed);
42. Throttle/power lever position;
43. Additional engine parameters;
44. Traffic alert and collision avoidance system;
45. DME 1 and 2 distances;
46. Nav 1 and 2 selected frequency;
47. Selected barometric setting (when an information source is installed);
48. Selected altitude (when an information source is installed);
49. Selected speed (when an information source is installed);
50. Selected mach (when an information source is installed);
51. Selected vertical speed (when an information source is installed);
52. Selected heading (when an information source is installed);
53. Selected flight path (when an information source is installed);
54. Selected decision height (when an information source is installed);
55. Electronic Flight Instrument System (EFIS) display format;
56. Multi-function/engine/alerts display format;
57. Thrust command (when an information source is installed);
58. Thrust target (when an information source is installed);
59. Fuel quantity in center of gravity (CG) trim tank (when an information source is installed);
60. Primary navigation system reference;
61. Icing (when an information source is installed);
62. Engine warning each engine vibration (when an information source is installed);
63. Engine warning each engine over temp. (when an information source is installed);
64. Engine warning each engine oil pressure low (when an information source is installed);
65. Engine warning each engine over speed (when an information source is installed;
66. Yaw trim surface position;
67. Roll trim surface position;
68. Brake pressure (selected system);
69. Brake pedal application (left and right);

70. Yaw or sideslip angle (when an information source is installed);
71. Engine bleed valve position (when an information source is installed);
72. De-icing or anti-icing system selection (when an information source is installed);
73. Computed center of gravity (when an information source is installed);
74. AC electrical bus status;
75. DC electrical bus status;
76. APU bleed valve position (when an information source is installed);
77. Hydraulic pressure (each system);
78. Loss of cabin pressure;
79. Computer failure;
80. Heads-up display (when an information source is installed);
81. Para-visual display (when an information source is installed);
82. Cockpit trim control input position – pitch;
83. Cockpit trim control input position – roll;
84. Cockpit trim control input position – yaw;
85. Trailing edge flap and cockpit flap control position;
86. Leading edge flap and cockpit flap control position;
87. Ground spoiler position and speed brake selection; and
88. All cockpit flight control input forces (control wheel, control column, and rudder pedal).

Typically, the sensors measure these parameters a few times per second. The recorder is also fitted with a "Pinger" that is activated when the recorder is submerged in water. The Pinger emits an acoustical signal that aids in the recovery of the device in case it falls into a body of water. Some physics specifications and requirements for the FDR include [NTSB, n.d.]:

1. Records up to 25 hours continuously;
2. Has an impact tolerance of 3400G's per 6.5 ms;
3. Underwater beacon (Pinger) operates for 30 days (when on, and has a shelf life of six years);
4. Is fire resistant up to 1100°C for at least 30 minutes;
5. Can be submerged up to 20,000 ft.

1.3.3 The Cockpit Voice Recorder (CVR)

The CVR is the device that records audio signals in the cockpit as well as the pilot's communication channel. The cockpit microphones, typically located above and between the pilot's seats, capture audio in the cockpit. Engine noises, instrument warning sounds, and other clicks and pops are all of interest to investigators. These sounds can be extremely useful in determining the cause of an accident. Furthermore, the crew's communication channel with ground stations (airport towers, air traffic control) is monitored and recorded. Most commercial CVR record four separate audio channels, FAA requires that fixed-wing CVR systems must record the following sources on separate dedicated channels:

1. First pilot station audio;
2. Second pilot station audio;
3. Cockpit area microphone;
4. Passenger loudspeaker system and any third and fourth crewmember stations audio.

The following are some physical specifications and requirements for the CVR [NTSB, n.d.]:

1. 30 minutes of continuous recording (up to two hours for newer models);
2. Has an impact tolerance of 3400 Gs / 6.5 ms;
3. Is fire resistant up to 1100°C for at least 30 minutes;
4. Can be submerged up to 20,000 ft;
5. Underwater beacon (Pinger) 37.5 kHz; operates for 30 days (when on, and has a shelf life of six years);
6. Four Channels.

Each recorder has an underwater locator beacon (ULB) to aid in finding it in the case of an overwater mishap. When the recorder is submerged in water, a device known as a "pinger" activates. It emits an acoustic signal at a frequency of 37.5 kHz, which a specialized receiver can pick up. The beacon has a range of 14,000 feet [NTSB, n.d.].

1.3.4 Other Types of Recorders

The aviation industry has developed various types of recorders to improve flight safety and aid in accident investigation. In addition to the CDR and FDR, there are other recorders that provide additional information about an aircraft's flight and crew activity.

1.3.4.1 Deployable Recorders

Deployable recorders are a type of recorder that combines the CVR, FDR, and an emergency locator transmitter into a single unit that is automatically deployed from the aircraft at the start of an accident sequence. The purpose of this recorder is to ensure that the recorder's data is preserved even if the aircraft is destroyed during an accident. The deployable unit has the capability of rapidly deploying and establishing a flight trajectory that clears the airframe during the accident sequence. The deployable unit is intended to float on water after deployment.

After deployment, the deployable unit begins transmitting an emergency signal that satellites and search aircraft/ships can detect. Deployable recorders are mostly found on water-based helicopters and military aircraft but are not installed in commercial planes [ATSB, 2014].

1.3.4.2 Combined Recorders

Combined recorders are another type of recorder that combines the functions of the CVR and the FDR into a single unit. When using combined recorders on an airplane, two units are required. One is located near the cockpit, and the other is toward the aircraft's rear (tail). The front-mounted recorder has the advantage of shorter cable distances between the cockpit area and the recorder, which reduces the possibility of the wires being breached during an in-flight fire or breakup. Impact survivability is maximized by traditional rear-mounted recorders. The purpose of this recorder is to reduce weight and space requirements on aircraft.

1.3.4.3 Image Recorders

Image recorders are a type of recorder that record images of all flight crew work areas including instruments and controls. The image recorder supplements existing information recorded by the CVR and the FDR. A general view of the cockpit area, instruments, and control panel displays provides insight into the cockpit environment, the serviceability of displays and instruments, crew activity, and the human/machine interface. However, image recorders are not widely adopted due to crew privacy issues and are not installed in commercial airliners. The purpose of this recorder is to provide additional information to help determine the cause of an accident.

1.4 Real-Time Data Transmission as a Solution

The problem that needs to be addressed is the storage of data collected by FDRs and CVRs. While these recorders are crucial in investigating aircraft accidents, they are merely storage devices that collect data from FDAUs or flight data acquisition and management systems (FDAMS). This book proposes an alternative method of storing this data by transmitting it in real time from the aircraft to a ground station. This will ensure that the data is available for analysis immediately, rather than waiting for the recovery of the physical recorder.

To achieve this goal, there are several objectives that need to be met. First, all aircraft should be able to transmit data to the ground while in flight. Second, the transmission of data should be done in a timely and organized manner, ensuring that all necessary information is received by the ground stations. Third, transmission protocols must prioritize aircraft in emergencies to ensure that critical data is sent as quickly as possible. Fourth, the system must continuously maintain a minimum backlog of information on board at all times. Finally, the system should use the least possible resources, such as available channels and bandwidth, to ensure efficient data transfer.

By addressing these objectives, the proposed real-time data transmission system can provide a more efficient and reliable way of storing and analyzing aircraft data, which can lead to improved safety measures and accident prevention.

1.5 System Capacity Requirements

Each aircraft must transmit the entire amount of data produced by the FDAU. Older FDAUs generate data at a rate of 64 12-bit words per second, or 768 bits per second (bps) [Matale, 2010]. Today's FDAUs generate up to 2048 12-bit words per second; for example, Airbus 380 uses 1024 words per second or 12,288 bps. However, Zubairi [2019] shows that 1800 bps data per airplane is enough to transmit the 88 parameters mandated by the FAA. We will assume that this is the minimum required data rate for each flying commercial airplane to transmit to the ground base station. In addition, we will assume that the aircraft will transmit its CVR data when requested or in emergency cases. The number of airborne aircraft in the sky at peak operational periods can approach 5400 in densely populated and urban areas with considerable air traffic, such as the United States. This book demonstrates a connection between the number of aircraft transmitting data and the channels and resources that are available. In our study, we aim to identify the number of channels that could be supported using existing communication resources (such as Iridium Satellite and very high frequency communication). Considering the peak operational number of planes, the system will need a minimum capacity of 1800 bps/aircraft $\times$ 5400 aircraft = 9,720,000 bps. The systems will be evaluated based on their ability to offer the capacity to carry the expected load efficiently. Regardless of the system employed, the systems will be required to deliver all the information from all the planes from takeoff to landing. In order to achieve the objective, two methodologies will be examined with the aid of two distinct systems. The initial system, which will be scrutinized in Chapter 4, relies on satellite communication. The subsequent solution in Chapter 5 involves employing very high-frequency digital link (VDL), particularly VDL mode 4, to transmit vital flight data, and its capability will be evaluated.

1.6 Summary

In this chapter, we presented a thorough investigation of the issues with traditional FDR/CVR and demonstrated that there was a significant number of aircraft crashes over the past two decades where either one of the FDR/CVR recorders was missing or damaged. Even if it were found, the search process would be risky and expensive. We also demonstrate the benefits of the alternative real-time flight recorder and how it outperforms the conventional FDR/CVR flight recorder now in use. Finally, we demonstrate the real-time flight recorder's minimum requirements. Chapter 3 presents existing communication techniques and protocols used by commercial airplanes.

References

Associated Press. US company resumes search for missing flight MH370, January 2018. URL https://www.telegraph.co.uk/news/2018/01/07/us-company-resumes-search-missing-flight-mh370/.

ATSB. Black box flight recorders, 2014. URL https://www.atsb.gov.au/publications/2014/black-box-flight-recorders.

S. Denyer. Search for Malaysia Airlines Flight 370 finally called off with mystery unsolved, January 2017. URL https://www.washingtonpost.com/world/search-for-malaysia-airlines-mh370-finally-called-off-with-mystery-unsolved/2017/01/17/3662d778-dc84-11e6-ad42-f3375f271c9c_story.html.

M. Dub and J. Parizek. Evolution of flight data recorders. *Advances in Military Technology*, 13(1):95–106, May 2018. ISSN 1802-2308. doi: 10.3849/aimt.01226.

D. Engber. Who made that black box? April 2014. URL https://www.nytimes.com/2014/04/06/magazine/who-made-that-black-box.html.

S. Fisher. Father of the blackbox, April 2017. URL https://www.historynet.com/father-black-box/.

A. Gartland. MH370: Ocean infinity search ends amid calls for new disclosures and further investigation, June 2019. URL https://changingtimes.media/2018/06/09/mh370-ocean-infinity-search-ends-amid-calls-for-new-disclosures-and-further-investigation/.

R. Grossi. Aviation recorder overview. *National Transportation Safety Board*, 2006.

S. Hradecky. Accident: Aeroflot SU95 at Moscow on May 5th 2019, aircraft bursts into flames during rollout and burns down, May 2019. URL https://avherald.com/h?article=4c78f3e6.

C. MacLeod, M. Winter, and A. Gray. Beijing-bound flight from Malaysia missing, March 2014. URL https://www.usatoday.com/story/news/world/2014/03/07/malaysia-airlines-beijing-flight-missing/6187779/.

A. Marsh. The inventor of the black box was told to drop the idea and "get on with blowing up fuel tanks". *IEEE Spectrum*, June 2021. URL https://spectrum.ieee.org/the-inventor-of-the-black-box-was-told-to-drop-the-idea-and-get-on-with-blowing-up-fuel-tanks.

S. Matale. Ground-based black box system implementation using satellite and VHF data link networks. Master's thesis, The University of Mississippi, Oxford, USA, 2010.

National Transportation Safety. Aircraft Accident Investigation Report. PT. Lion Airlines Boeing 737 (MAX); PK-LQP Tanjung Karawang, West Java, Republic of Indonesia, October 2018.

NTSB. Cockpit Voice Recorders (CVR) and Flight Data Recorders (FDR), n.d. URL https://www.ntsb.gov/news/Pages/cvr_fdr.aspx.

SKYbrary. Flight Data Recorder (FDR), 2014. URL https://www.skybrary.aero/articles/flight-data-recorder-fdr.

P. Yeung. EgyptAir crash: Black box cockpit voice recorder from flight MS804 'found in Mediterranean Sea, June 2016. URL https://www.independent.co.uk/news/world/middle-east/egyptair-crash-flight-ms804-black-box-cockpit-voice-recorder-found-mediterranean-sea-latest-a7085611.html.

J. Zubairi. Your flight data is on us!! In *2019 IEEE 16th International Conference on Smart Cities: Improving Quality of Life Using ICT & IoT and AI (HONET-ICT)*, pages 241–243, 2019. doi: 10.1109/HONET.2019.8908019.

2

State of the Art

In traditional aviation accident investigations, flight data and cockpit voice recorders (FDR/CVR) have been critical resources for determining the cause of a disaster. However, the fact that they were physically attached to the aircraft posed a significant challenge. If the plane was lost, the data could not be retrieved. As highlighted in the introductory chapter, investigators have faced this issue repeatedly in numerous flight accidents.

The solution to this problem lies in streaming FDR/CVR data to the ground in real-time, periodically, or when the aircraft behaves abnormally. However, implementing this solution comes with its own set of challenges, including cost, privacy concerns, and a lack of regulatory framework.

This chapter focuses on reviewing the works of researchers and industry specialists who have proposed, designed, and tested different systems for transmitting real-time flight data from aircraft to ground stations. It discusses the challenges associated with implementing such systems and presents available products in the market.

In Chapters 4–6, proposed solutions to the challenges faced by these available products will be addressed. The feasibility and implementation of real-time flight data transmission systems have been explored in various studies. This chapter will review and discuss the findings of these studies to provide insights into the possibilities and challenges of real-time flight data transmission.

Various studies have been conducted to explore the feasibility and implementation of real-time flight data transmission systems.

2.1 Preceding Research

Making the instant FDR/CVR data available by transmitting flight data instantaneously to a ground base station has received great interest from researchers and

Real-Time Ground-Based Flight Data and Cockpit Voice Recorder: Implementation Scenarios and Feasibility Analysis, First Edition. Mustafa M. Matalgah and Mohammed Ali Alqodah.
© 2024 The Institute of Electrical and Electronics Engineers, Inc. Published 2024 by John Wiley & Sons, Inc.

manufacturers and examined by industry specialists as an alternative to the traditional methods during the last two decades, especially after the Malaysian MH370 flight disaster in 2014 [Schoberg, 2003, Werfelman, 2009, Matale, 2010, Kavi, 2010, Zubairi and Er, 2012, Zubairi, 2017, 2019, Shaikh et al., 2019, Wang et al., 2019, FLYHT, 2021]. This will ensure that all the information required to investigate airplane crashes is available instantly to the crash investigators. The author of Schoberg [2003] investigated the feasibility of a system that transmits flight data from an aircraft to a ground recording station in a thesis published in 2003 and discussed the required technical framework. The requirements for information flow security and assurance to ensure data confidentiality, integrity, availability, and authenticity were explored. According to the author, the real-time flight data transmission system was unlikely to be implemented at that time due to existing aviation wireless telecommunication technologies then. Implementing such a system required the development of airborne collection computers, additional aircraft systems to route sensor data to the computers, possibly additional radios, a sophisticated data network with large capacity, large data warehouse computer systems, and a system to examine the stored data. The author made the point that a smooth data link handoff needs more study and development to preserve the secure connection between the aircraft and the ground base station across different communication platforms. According to an article published in Aero Safety World Magazine in August 2009 [Werfelman, 2009], a potential alternate transfer of flight data to the ground station is linked to technological advancements in data transmission between the air and the ground. The author makes some suggestions for how to improve the existing traditional black boxes, which she referred to as "fixed black boxes." One option is to use military aircraft technology to send the location of the black box immediately after the crash incident when it gets detached. The author also proposed combining the separate FDR and CVR into a single unit. In the same article [Werfelman, 2009], it was indicated that after the Air France Flight 447 crash [BEA, 2009] both Airbus and Boeing studied the use of other alternatives for reinforcing flight data recovery capabilities including examining the feasibility of extended data transmission. However, they pointed out some issues associated with transmitting flight data to a ground station, such as the enormous amount of data to be transmitted and the difficulty in maintaining a reliable link between aircraft and satellites, especially when the aircraft have unusual altitudes. In a 2010 thesis [Matale, 2010], the author demonstrated different scenarios on how real-time flight information data could be transmitted from aircraft to ground stations through reliable links via more advanced wireless telecommunication technologies, creating real-time ground-based FDR/CVR recorders. In addition, the feasibility of the researched communication infrastructures for complete data transfer from aircraft to storage and analysis facilities on the ground with a tolerable delay was also assessed by the author. Two main infrastructures were examined and explored, the very high-frequency digital link (VDL) and the Iridium satellite system. The author demonstrated quantitative case studies of resources required to enable data transfer from specific numbers of aircraft based in a high-traffic area, namely the United States. In Kavi [2010], the author discussed an idea for ground-based real-time flight data recording; he named it a "glass-box system" in which data from the aircraft would be transmitted in real-time to ground stations for analysis on the spot or later. The author suggested that the transmission would use high-bandwidth radio unless the plane was flying over water, in which case lower-band satellite links would be used. Consequently, investigators could thus look for symptomatic patterns in data from the crashed aircraft to determine what happened to it more precisely. In Zubairi and Er [2012] and Zubairi [2017], the authors proposed and built a design for a distributed, scalable, and fault-tolerant flight tracker system. In this system, a protocol comprised of a set of algorithms was developed for the reliable transmission of real-time flight data to distributed ground servers. This protocol makes use of existing air and ground aviation infrastructure to ensure adaptability; however, this protocol is limited to local areas that include these specific distributed ground servers and seems not to utilize global communication technologies for intercontinental international flights that will require continuous connectivity throughout their trips. In Zubairi [2019], the author describes the operation of the flight data tracker system (FDT). It operates on the plane server, as well as the central and data servers. The FDT system sends flight data to a central server at the origin airport or in the cloud. The shortest radio path between the origin and destination airports is determined by the central server. It displays a list of cloud data servers linked to radio towers along the route to the destination. The plane server initiates data transmission after establishing a handshake with the first data server. Following a handshake, each data server receives the data and forwards it to the central server at the origin or in the cloud. When the flight is completed, the central server has the complete flight data. In Shaikh et al. [2019], the authors examined the most recent developments in emerging network communication systems for commercial aviation and addressed their security issues. According to the authors, Airlines are quickly gathering aircraft's vital data and using it to improve their operations through predictive analysis. It is critical to transfer the majority of collected data to prominent terrestrial data center locations as part of this process. This is where big data analytics and cloud computing come into play, providing near real-time (if not real-time) situational awareness as well as significantly improved decision support and resource efficiency. An aircraft, for example, could continuously transmit black box data to aid in real-time route optimization, identifying potential faults, and improving flight safety. As a result, many aircraft manufacturers are already employing a wide range of sensors to collect critical data and perform (off-line) machine learning analysis and optimization of flight routes, fuel costs, waiting times, take-off and landing schedules, and so on. In Wang et al. [2019], the authors propose a software-defined wireless networking framework to fully exploit inter-aircraft air-to-air radio links in order to stream flight data from the aircraft to a ground control center in real-time. A time-expanded connectivity graph for the network controller was developed to maintain a holistic view of the time-varying network status and propose a branch and price algorithm to optimize the flight data flows transmitted via air-to-air links.

2.2   Wireless FDR/CVR Products in Market

There is a dearth of wireless FDR/CVR systems on the market. To our best knowledge, only two products are currently available, the Honeywell Connected Recorder and the FLYHTStream AFIRS 228 BlackBox streaming system. In the following, we will describe each of these two products along with their specifications.

2.2.1   Honeywell Connected Recorder

Honeywell has been a supplier of traditional black boxes to the aviation sector for more than 60 years. With the release of its new Connected Recorder-25, the company is investigating the potential of wireless black box transmissions. Honeywell Aerospace and Curtiss-Wright Corporation have worked together for the past few years to reinvent the aircraft CVR and FDR by utilizing the in-flight connection. They have collaborated on the development of the hardware for the "next generation" black box, also known as the (wireless FDR/CVR) Product known as HCR 25, with Honeywell developing software features for quicker access to real-time data [Nott, 2019]. These new voice and flight data recording options will help aircraft owners, operators, and manufacturers decrease aircraft downtime. Moreover, in the unlikely event of an emergency, it helps with the subsequent investigation via the prerecorded information about the flight status that is available in the ground base stations. This HCR 25 operates via current and future global satellite networks, communicating with a dedicated Honeywell data center. This new product of CVR/FDR gives owners, operators, and manufacturers the option to access the data at any time from anywhere, which results in the potential for better maintenance, predictability, and operational insight through data analytics, not only that but also in the event of an emergency, the data on board will be quickly accessible to authority personnel for further investigations [Nott, 2019].

Honeywell is also collaborating with satellite operator Inmarsat to ensure its "black box in the cloud" solution delivers on its promises to the commercial airline, cargo transport, and business jet markets. Inmarsat's SwiftBroadband-based cockpit communications solution SB-Safety (SB-S) – which runs over the company's L-band satellites, via onboard avionics hardware manufactured by Honeywell or Cobham – is a natural conduit for the "black box in the cloud" application because it is safety services-approved. In addition, Airbus already offers SB-S as a preferred line-fit service on various aircraft types [Kirby, 2019]. Depending on Inmarsat satellite systems, the Honeywell recorder is anticipated to be a costly option. Later in Chapter 5, we will discuss alternative collaborative solutions.

2.2.1.1 Honeywell Connected Recorder (HCR-25) Specifications

The HCR-25 is a recording system for airplanes that enables the recording of up to 25 hours of parametric flight data, four channels of cockpit voice recording, and Controller-pilot data link communication (CPDLC) recording, respectively. In addition, the HCR-25 follows the guidelines laid out by RTCA (Radio Technical Commission for Aeronautics) specifications DO-178C and DO-254 regarding its design. It satisfies all of the Federal Aviation Administration (FAA)'s and EUROCAE's existing and upcoming criteria. It comes equipped with an ultrasonic locator beacon (ULB) that can transmit for ninety days. In addition to satisfying regulations, HCR-25 operators can reap the benefits of real-time data streaming and cloud-upload capabilities, which are made possible by Honeywell's Connected Aircraft software. These capabilities make it possible for operators to quickly and remotely retrieve data from the aircraft to store or analyze it. Figure 2.1 provides an illustrative depiction of the operational mechanisms inherent in the Honeywell black box system. In parallel, Figure 2.2 depicts the Honeywell HCR-25 recorder products [Honeywell, 2022].

2.2.2 FLYHTStream

The FLYHTStream system is a cutting-edge, market-leading innovation that sends instantaneous alerts and streams black box data in the case of an emergency on board an airplane. Automatic activation of FLYHTStream can be performed in flight by a set of predetermined variables selected by the pilots or on the ground by airline operations. It interprets alarms and messages from the AFIRS system and sends them to key stakeholders on the ground, like the airline, operation centers, and regulators. The raw FDR data are transformed by animation software into visual data that can be seen from any computer, giving ground workers a visual of the controls and a knowledge of what is happening onboard the aircraft.

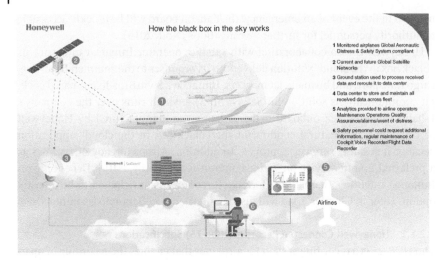

Figure 2.1   Honeywell Aerospace and its partners are readying for real-time black box transmissions. Source: Honeywell International Inc.

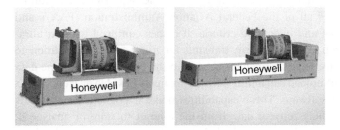

Figure 2.2   Honeywell Recorder (HCR-25). Source: Honeywell International Inc.

Patents for this data-streaming system have been granted to FLYHT in Canada, the United States, and China, with applications pending in other countries.

FLYHTStream is a revolutionary, industry-leading technology that performs real-time triggered alerting and black box data streaming in the event of an abnormal situation on an aircraft. FLYHTStream can be activated automatically by a set of predetermined factors by the pilots or on the ground by airline operations. It uses the AFIRS system's onboard logic and processing capabilities in combination with ground-based servers to interpret and route alerts and messages to key groups on the ground, such as the airline, operation centers, and regulators. Animation software converts the raw FDR data into visual data that can be viewed from any computer, providing ground personnel a view of the controls and awareness of what is happening onboard the aircraft. FLYHT has been awarded

Canadian, United States, and Chinese patents for this data-streaming technology, pending in other countries [FLYHT, 2021]. FLYHT AFIRS relies on the pricey Iridium satellite network. In Sections 6.1 and 6.2, we will discuss implementing collaborative aviation networks at a lower cost.

2.2.2.1 FLYHT AFIRS 228 Family Specifications

Automated Flight Information Reporting Systems 228 family in Figure 2.3 provides Iridium Global Voice and Data service in real-time. The AFIRS 228 Iridium SATCOM family offers a direct communication link to your aircraft, ensuring safe, secure operations, and greater in-flight operational control. AFIRS 228™ models are designed and engineered to meet various aviation requirements. These provide the crew with reliable voice and data services, including SATCOM voice, global flight tracking, two-way text messaging, aircraft health monitoring including engine trending, engine/airframe exceedances, real-time engine data analytics, fuel management, real-time flight data management, and Aircraft Communications Addressing and Reporting System (ACARS) over Iridium, as well as live black box streaming capabilities [FLYHT, 2021].

In order to address the issue at hand, a viable solution is to transmit FDR/CVR data via streaming to the ground. The streaming may occur in real-time, periodically, or triggered by abnormal aircraft behavior. However, this approach is not without challenges, including cost, privacy concerns, and a lack of regulatory framework. Section 2.3 will focus on outlining the primary issues confronting Wireless FDR/CVR products.

Figure 2.3 FLYHT AFIRS 228. Source: FLYHT [2021] Aerospace Solutions Ltd.

2.3 Wireless FDR/CVR Challenges

Despite its immense value, commercial production of wireless FDR/CVR recorders must address four key challenges.

2.3.1 The Cost Aspect

While FDR/CVR data streaming has broad support, most airlines view its integration as too expensive. The airline industry is understandably wary of adding expenses that could further eat into profits. There is a very slim chance of making any money in this field. The cost of fuel and government fees contribute to airlines low-profit margins of barely 1%, and in 2012 they produced earnings of only US $4 for every carried passenger, as reported by The Economist magazine [W, 2014]. Travel companies claim that implementing a wireless recorders system that streams data in real-time will increase the price of every flight. The airlines claim that the flight is unprofitable if real-time FDR/CVR data transmission is implemented [Mayer, 2014].

This misconception is based on a misunderstanding of what can be accomplished with current technology. For instance, the FLYHTStream AFIRS (Automated Flight Information Reporting Systems) (Irridum satellite-based system), one of only two available products, does not continuously transmit flight data from black boxes, it transmits data only during abnormal events. As data are only streamed after an abnormal event, satellite transmission costs are drastically reduced. According to Iridium's pricing, sending FDR/CVR data via their satellites to the ground would set you back between US $5 and US $7 per minute. If this technology had been on the missing Malaysian Airlines flight and live-streamed for the estimated seven hours following the first sign of trouble, the cost would have been approximately US $3,000 [Mayer, 2014]. Considering the amount of time, money, and effort spent on the fruitless search for MH370 recorders, the expense of operating a live-streaming equivalence appears insignificant. This expenditure is not excessive when considering the potential outcomes. Not to mention AFIRS is based on a costly Iridium satellite system. Later in Chapter 5, we will address less costly systems based on collaborative airspace networks via VDL communications.

2.3.2 Industry Factors

New commercial airplanes are built to last for decades, so updating their technology can be time-consuming and expensive. The cost to retrofit existing fleets is likewise high. Integrating new technology into older planes that were built with a high degree of precision by companies like Boeing and Airbus is a challenge. Similar to other "long-cycle" industries, aviation experiences very slowly emerging rates of development [Singh, 2022]. Unlike, for example, the mobile cellular wireless communication systems industry, which emerges from one generation to a new generation every decade while it could take three to five decades for the aviation industry to do so.

2.3.3 Lack of Regulations

There is currently no regulatory requirement that new aircraft models include live-streaming black box data, and any potential modifications would be slow and complicated to implement. For the past two decades, the National Transportation Safety Board (NTSB) has been urging the installation of video recording equipment in cockpits. Still, pilot unions have fought against the initiatives owing to privacy concerns. They have also expressed opposition to the notion of making their voice recordings accessible to anyone other than those responsible for investigating the accident [Minkoff, 2021]. Currently, to our best knowledge, there are no worldwide ongoing efforts by the aviation regulatory authorities to develop regulations regarding ground-based wireless FDR/CVR products. This could be driven if more and more products of this kind are available in the market.

2.4   Summary

In this chapter, we covered a comprehensive overview of the research that has been conducted on real-time flight data transmission systems. The feasibility of such systems has been explored by researchers and industry for over two decades. The focus on these systems has increased since the Malaysian MH370 flight disaster in 2014. Researchers have explored different communication infrastructures such as the VDL and the Iridium satellite systems. The potential alternate transfer of flight data to the ground station has been linked to technological advancements in data transmission between the air and the ground. Several proposals have been made, including transmitting flight data in real-time to ground stations for analysis on the spot or later. Different methods have been developed for the reliable transmission of real-time flight data to distributed ground servers. However, some limitations such as the enormous amount of data to be transmitted, the difficulty in maintaining a reliable link between aircraft and satellites, especially when the aircraft has unusual altitudes, and the limited global communication technologies for intercontinental international flights have been identified. We also discussed the issues the wireless FDR/CVR recorders face in the market, particularly the cost factor. We highlighted the two available wireless FDR/CVR in the market and demonstrated that they are pricey since they rely on satellite communication. In Chapters 4 and 5, we will cover the satellite implementation of wireless BlackBox as well as an implementation that relies on VDL and a cooperative approach that reduces the total cost.

References

BEA. Interim report on the accident on 1st June 2009 to the Airbus A330-203 registered F-GZCP operated by Air France Flight AF 447 Rio de Janeiro –Paris, 2009.

FLYHT. Real-Time Global Communications-AFIRS 228™, 2021. URL https://flyht.com/airborne-hardware/afirs-228/.

Honeywell. Honeywell Connected Recorder-25, 2022. URL https://aerospace.honeywell.com/us/en/products-and-services/product/hardware-and-systems/recorders-and-transmitters/flight-recorders.

K. Kavi. Beyond the black box. *IEEE Spectrum*, 47(8):46–51, 2010. doi: 10.1109/MSPEC.2010.5520630.

M. Kirby. Honeywell works with Inmarsat to make black box in the cloud a reality, March 2019. URL https://runwaygirlnetwork.com/2019/03/honeywell-works-with-inmarsat-to-make-black-box-in-the-cloud-a-reality/.

S. Matale. Ground-based black box system implementation using satellite and VHF data link networks. Master's thesis, The University of Mississippi, Oxford, USA, 2010.

A. Mayer. AirAsia flight QZ8501: Why airlines don't live-stream black box data, March 2014. URL https://www.cbc.ca/news/science/airasia-flight-qz8501-why-airlines-don-t-live-stream-black-box-data-1.2586966.

Y. Minkoff. Why don't airlines live-stream black box data? January 2021. URL https://seekingalpha.com/news/3650388-why-dont-airlines-live-stream-black-box-data.

G. Nott. Black box recorders reinvented to live-stream data, February 2019. URL https://www2.computerworld.com.au/article/657699/black-box-recorders-reinvented-live-stream-data/.

P. Schoberg. Secure ground-based remote recording and archiving of aircraft "Black Box" data. Master's thesis, Naval Postgraduate School, California, USA, 2003.

F. Shaikh, M. Rahouti, N. Ghani, K. Xiong, E. Bou-Harb, and J. Haque. A review of recent advances and security challenges in emerging E-enabled aircraft systems. *IEEE Access*, 7:63164–63180, 2019. doi: 10.1109/ACCESS.2019.2916617.

S. Singh. Why don't airlines live stream black box data? April 2022. URL https://simpleflying.com/why-dont-airlines-live-stream-black-box-data/.

S. W. Why airlines make such meagre profits, February 2014. URL https://www .economist.com/the-economist-explains/2014/02/23/why-airlines-make-such-meagre-profits.

Y. Wang, K. Wan, C. Zhang, X. Zhang, and M. Pan. Optimized real-time flight data streaming via air-to-air links for civil aviation. In *ICC 2019 - 2019 IEEE International Conference on Communications (ICC)*, pages 1–6, 2019. doi: 10.1109/ICC.2019.8761955.

L. Werfelman. Thinking outside the black box. *Aero Safety World Magazine*, pages 24–27, August 2009.

J. Zubairi. Flight data tracker, August 2017. US Patent 9,718,557.

J. Zubairi. Your flight data is on us!! In *2019 IEEE 16th International Conference on Smart Cities: Improving Quality of Life Using ICT & IoT and AI (HONET-ICT)*, pages 241–243, 2019. doi: 10.1109/HONET.2019.8908019.

J. Zubairi and A. Er. Fault tolerant aviation data tracker design. In *High Capacity Optical Networks and Emerging/Enabling Technologies*, pages 047–051, 2012. doi: 10.1109/HONET.2012.6421433.

3

Aviation Communication Overview

During the early days of aviation, it was assumed that skies were too big and empty to the extent that it was almost impossible that two planes would collide. However, in 1956 a famous crash occurred over the Grand Canyon, which sparked the Federal Aviation Administration (FAA) creation. Moreover, due to the fact that aviation was roaring during the Jet Age communication technologies development became vital. This was initially seen as a very difficult task; ground control stations in the early days used visual aids to provide signals to pilots in the air. With the advent of portable radios small enough to be placed in planes, pilots were able to communicate with people on the ground. With later developments, pilots were then able to converse not only air-to-ground (A-G) but also air-to-air (A-A). Today, aviation communication relies heavily on the use of many systems where planes are outfitted with the newest radio and Global Positioning System (GPS) systems, as well as Internet and video capabilities.

3.1 History

During the last more than a century, following the first flight ever (in 1903), air travel has emerged as a crucial means of transportation not only for people but also for cargo, which created a revolution in the way people transport. This airline revolution created a new business trend and economy [Editorial Team, 2023]. Since communication between the pilots and ground station controllers has to be an integrated component in aviation for safety purposes, in the early days this was accomplished by visual signaling using, for instance, colored paddles or hand signs as shown in Figure 3.1. This means was dedicated to the ground crew to guide the pilot while landing or departing but did not apply to the pilots [Mahmoud et al., 2014]. Since the flight revolution started shortly

Real-Time Ground-Based Flight Data and Cockpit Voice Recorder: Implementation Scenarios and Feasibility Analysis, First Edition. Mustafa M. Matalgah and Mohammed Ali Alqodah.
© 2024 The Institute of Electrical and Electronics Engineers, Inc. Published 2024 by John Wiley & Sons, Inc.

Figure 3.1 The earliest communication with aircraft was by visual signaling. Farman [2023]. Source: Unknown Source/Wikipedia Commons/Public Domain.

before the world war era, pre, during, and post World War I ignited a great deal in flight development and aviation. In the following, we will discuss aviation communication evolution during the last century.

3.1.1 Wireless Telegraphy Era

In 1911, wireless telegraphy started to be deployed in aviation and in particular was used in the Italo-Turkish War in 1911 and 1912. The Royal Flying Corps, the air arm of the British army before and during the first world war, began experimenting with "wireless telegraphy" in aircraft and started radio experiments later in 1913, successfully improving wireless radio's performance. Even though wireless communication systems in aircraft remained experimental, a successful practical prototype remained undeveloped for many years. Due to the lack of advanced technologies at that time, radios were heavy and unreliable. Additionally, the absence of security technology at that time, radio communication was rarely used by ground forces as signals are easily intercepted by pausing forces; consequently, during the world war I era, soldiers used large panel cutouts to distinguish friendly forces instead of using wireless technologies, they were also used as directional devices to help pilots navigate back to friendly and familiar airfields.

3.1.2 Analog Radio Communication Era

The first time speech had ever been communicated between an airplane and ground personnel was accomplished in April 1915. Charles Prince "A Marconi company in Britain engineer" created the first aviation tube receiver, and Captain J. M. Furnival of the Royal Air Force became the first person to hear a voice broadcast from the ground [Flying the Beams, 2020]; later in June, the 20-mile air to ground voice transmission took place at Brooklands, United Kingdom, on the other hand, ground to air two-way voice communication was available and installed by July 1915 after it was initially achieved using Morse codes. After that, particularly in early 1916, the Marconi company started to prototype and production of air to ground transmitters and receivers that were used in the war over France at the time. The first air-to-ground radio in the USA with a range of 2000 yards was developed in 1917, in which the U.S. Army Signal Corps directed AT&T to develop such a transmitter. By 4 July of the same year, it was able to set up a two-way communication between pilots and ground staff.

Following the First World War, there were several advancements in the development of radio communication systems. These developments were aimed at enhancing range and performance. In 1915, radios had a limited range of only two miles. However, in 1917, radiotelephone sets were manufactured and put into use in aircraft and ground stations. These sets could reach other aircraft and ground stations at distances of 25 and 45 miles, respectively [Higginbotham, 2018]. However, it was not until 1930 that airborne radios became reliable enough and had sufficient power to be effective. That same year, the International Commission for Aerial Navigation mandated that all aircraft carrying 10 or more passengers should be equipped with wireless equipment.

In the mid-1930s, the development of radar marked a significant breakthrough in air-to-ground communication. Radar systems could track planes in the air, providing information about their distance, direction, speed, and even aircraft type. This advancement improved air traffic control (ATC) and navigation aids for pilots. After several years of testing and modification, the US Civil Aeronautics Administration launched radar departure control procedures at the Washington National Airport on 7 January 1952. Subsequently, they deployed radar approach control procedures at the airport six months later.

One of the most notable instances that demonstrated the importance of radio communication in aviation was the Grand Canyon disaster of 1956, where two commercial planes collided over the Grand Canyon, resulting in the death of all 128 passengers on board. The collision occurred in uncontrolled airspace, highlighting the need for effective radio communication and the responsibility of pilots to maintain separation. Following the tragedy, the United States government implemented several changes to improve aviation safety, including the requirement for all aircraft to be equipped with two-way radios. This change has contributed to preventing similar incidents in the future.

Since the early days, radio technology has progressed significantly, and it now plays a crucial role in modern aviation. Radios are used for various purposes, including navigation and ATC, which is essential for safe flying.

In the years that followed these legislations by the US government to increase air safety, the very high frequency (VHF) band was mainly used for radiotelephone between pilots and ground station controllers personnel. Even though the technologies used have been continuously evolving since then, the main principle is still the same today. Initially, VHF-reserved bandwidth implementations (also known as aircraft band or airband) were based on 140 channels with a spacing of 100 kHz in a spectrum range from 118 to 137 MHz. This spectrum splitting into sub-channels with spacing is performed to ensure the efficient sharing of resources. Later, during the era of 1979–1989, the aircraft band splitting was modified to include 760 channels with 25 kHz spacings.

3.1.3 Digital Radio Communication Era

Following the digital radio communication breakthrough that started in mid-1990, aviation communication adopted this technology as well, which significantly increased capacity by reducing the bandwidth requirements for digital speech transmission (using compression techniques and vocoders[1]). Consequently, this facilitates splitting the airband into multifold numbers of channels (2280 channels with a spacing of 8.33 kHz). Moreover, to enhance reliability and extend availability, the aircraft band was expanded to include high frequency (from 3 to 30 MHz) digital voice radios and VHF. In addition, aviation radios have been further augmented with Satellite communications (SATCOM) when the pilot is out of range of direct communication with VHF/HF ground stations.

3.1.4 Digital Data Link Era

Over the past 50 years, the significant growth in air traffic has resulted in airspace becoming increasingly congested, leading to limited availability of radio communication resources [Mahmoud et al., 2014]. To tackle this problem, a digital data link scheme was proposed as an alternative solution. This system enables the transmission of short digital messages between aircraft and base stations using VHF/HF or SATCOM. Aircraft Communications Addressing and Reporting System (ACARS) was the first data link means introduced in the 1970s by Aeronautical Radio Incorporated (ARINC) to reduce crew workload and improve

---

1 A vocoder is a type of speech coding chip that analyzes and synthesizes human voice signals for the purposes of audio data compression, multiplexing, and voice encryption.

data integrity. The system was initially used for digital voice communication using VHF channels and has since been extended to include SATCOM and high frequency (HF) links.

As air traffic continued to grow, integrating different digital data link technologies became necessary. This integration, along with its applications to air traffic management (ATM) air–ground communications, forms part of the larger set known as communication, navigation, and surveillance/air traffic management (CNS/ATM) systems. The International Civil Aviation Organization (ICAO) established a special committee called the future air navigation system (FANS) in the early 1980s to identify and evaluate new concepts and technologies that could benefit international civil aviation. The goal was to define operational concepts for future ATM using digital technologies. Boeing and Airbus developed FANS-1 and FANS-A based on the FANS committee's recommendations, which were eventually combined as FANS-1/A.

For data link operation, an aircraft is considered FANS-1/A equipped if it has the air traffic services facilities notification (AFN) capabilities. The use of ACARS and FANS-1/A has significantly enhanced the safety and efficiency of ATM by improving communication between controllers and pilots. Today, air–ground communications may be based on analog voice, Plain Old Aircraft Communications Addressing and Reporting System (POA), FANS-1/A, or even FANS 2/B with the latest improvements. These successive steps have resulted in a heterogeneous and relatively complex aeronautical world.

3.2 Communication Traffic Classes

In the context of aeronautical telecommunications, the ICAO has established four communication traffic classes, which are outlined in Annex 10 (Volume 3, Chapter 3) of the International Standards and Recommended Practices and Procedures for Air Navigation Services. These classes are air traffic service communications (ATSCs), aeronautical operational control (AOC) communications, airline administration communications (AACs), and Aeronautical passenger communications (APCs) (Figure 3.2). These four categories of communication are described by Mahmoud et al. [2014] in more detail as follows:

1. ATSCs are critical and include communication between the pilot and ATC to ensure safe, efficient, and speedy flight. The services in this class can be supported by voice broadcast or data communications and may include meteorological and route information.
2. AOC communications are also critical and encompass communication required for exercising authority over flight initiation, continuation, diversion, or termination for safety, regularity, and efficiency reasons. Examples of this

Figure 3.2 Aeronautical communication traffic classes. Source: Mahmoud et al. [2014]/John Wiley & Sons.

class include communication between airline companies and their aircraft regarding maintenance messages, fuel levels, departure, and arrival times.

3. AACs are noncritical and include communications necessary for exchanging aeronautical administrative messages. These messages are not linked to the security or efficiency of the flight, and examples include information regarding passengers, special cleaning requests, and other non-essential communication.

4. APCs are also noncritical and include communication-related to nonsafety voice and data services to passengers and crew members for personal communication. These services include in-flight entertainment and Wi-Fi access. The classification of each communication traffic class helps to define specific requirements and properties for each application, depending on the class to which it belongs.

The term "critical" above was noted by Mahmoud et al. [2014] as that

critical communications follow specific international rules defined by ICAO (for example, only some dedicated frequency bands can be used), particularly for ATSC and AOC and are based on dedicated systems. The latter must meet the stringent quality of service (QoS) requirements, mainly based on transaction time, continuity, availability, and integrity parameters. These regulatory constraints do not apply to non-critical communications even if they may have to meet some requirements according to the applications (e.g. delay for passenger telephony). Sometimes, AAC services are included in AOC services that give only three classes: ATSC (or ATS) and AOC as safety-related classes, and APC as the non-safety class. The current solution in civil aviation communication to ensure segregation between safety service classes and APC is physical segregation between critical and non-critical communication. The pieces of equipment aboard aircraft are physically different.

3.3 Main Actors and Organizations

Several actors and organizations are involved in communication systems that are dedicated to datalink, service design, standardization, deployment, and maintenance. These actors and organizations can be classified into four main categories as follows [Mahmoud et al., 2014].

3.3.1 Aviation Authorities

The first category includes aviation authorities, whose primary objectives are to define principles and techniques for international air navigation, promote the planning and development of international air transport, and ensure safe and orderly growth. The ICAO is the most important organization in this category. It is a specialized agency of the United Nations that consists of 191 of the 193 UN members. The ICAO Council adopts standards and recommended practices concerning air navigation and its infrastructure, flight inspection, prevention of unlawful interference, and facilitation of border-crossing procedures for international civil aviation. The European Organisation for Civil Aviation Equipment (EUROCAE) and the Radio Technical Commission for Aeronautics (RTCA) are two well-known Standard-Developing Organizations (SDOs) in this category. EUROCAE certifies aviation electronics in Europe, while RTCA develops consensus-based recommendations regarding CNS/ATM system issues. The European Organisation for the Safety of Air Navigation (EUROCONTROL) is another important actor in this category. It was founded in 1960 and is an international organization working for seamless European ATM. EUROCONTROL coordinates and plans ATC for all of Europe, working in close partnership with several organizations such as national authorities, air navigation service providers (ANSPs), civil and military airspace users, and airports.

3.3.2 Air Transport Industry

The second category, known as the "Air Transport Industry," involves major plane manufacturers like Airbus and Boeing, as well as communication equipment suppliers. These manufacturers and suppliers play a critical role in the development of communication systems dedicated to data links as they are responsible for designing, producing, and installing specific equipment that complies with the regulations and guidelines set by aviation authorities.

3.3.3 Aviation Datalink Service Providers

The third category is the Aviation datalink service providers (DSPs) and communication message service providers who ensure the reliability and integrity of the transmission media and message communication between pilots and ground stations. These providers are responsible for the creation and management of multiple data links that transmit various messages related to specific applications between the aircraft and ground. Ground stations located at airports and other sites operate a large network that provides comprehensive VHF, HF, and SATCOM coverage in continental and oceanic airspaces. Société Internationale de Télécommunications Aéronautiques (SITA) and ARINC are two well-known DSPs that offer VHF data link (VDL) mode 2 services, while Inmarsat and Iridium provide SATCOM links for oceanic airspaces. ICAO has authorized only some satellite systems, including Aero-H/H+/I/L proposed by Inmarsat and Iridium. Inmarsat operates four geostationary satellites that cover about 97% of the earth's surface, while the Iridium system is based on a constellation of 66 cross-linked satellites and seven spares that create its network of global coverage.

3.3.4 Aviation Stakeholders

The last category of participants in the aviation industry are referred to as aviation users, which includes clients who utilize the services of all the other actors mentioned above. Aviation users can be classified into three main categories, as described below.

3.3.4.1 ANSPs

The first category of aviation users are ANSPs, responsible for ATC. ANSPs may be government agencies, state-owned entities, or private organizations such as Directorate General for Civil Aviation (DGAC) in France and the FAA in the United States. DSNA, which manages ATC communication and information in France, is another example of an ANSP. The FAA, as the national aviation authority of the United States, regulates and oversees all aspects of American civil aviation.

3.3.4.2 Airlines

The second category of aviation users are airlines, companies that offer air transportation services for passengers and freight. Examples of airlines include

Lufthansa, United Airlines, American Airlines, Royal Jordanian, Qatar Airways, Delta Airlines, Air France, and FedEx.

3.3.4.3 Meteorological Centers

The third category of users are the meteorological centers that collect meteorological information and produce forecasts. Meteorological information can also be collected by aircraft that used embedded equipment named data management unit (DMU) for acquiring data related to weather observations such as temperature, and wind at various positions in the sky. This information collection is performed by several airlines and sent to international meteorological centers as input to weather forecast models. More information on sky weather conditions, such as turbulences, can also be sent by the pilots during their flight to inform ATC stations. On the other hand, meteorological centers provide weather information and forecasts that can be sent to aircraft during the flight. These data are generally provided to the aircraft through particular data link applications. For instance, significant meteorological effects (SIGMETs) are advisories regarding significant meteorological conditions that could affect the flight. A meteorological aerodrome report (METAR) allows the pilot to get updated weather conditions and forecasts at the departure and destination airports and other airports along the route.

3.4 Spectrum Allocation to Aeronautical Services

Spectrum allocation to aeronautical services is the process of assigning radio frequencies and other electromagnetic spectrum resources to different types of aeronautical services, such as air traffic control, aircraft communication, navigation, and surveillance systems. It is a critical aspect of ensuring the safe and efficient operation of air travel. Effective spectrum allocation is essential for ensuring that these systems operate efficiently and without interference. For example, ATC systems rely on radio frequencies to communicate with pilots and coordinate air traffic. If these frequencies were allocated improperly, it could lead to communication breakdowns or other safety issues. Spectrum allocation for aeronautical services is regulated by national and international organizations, such as the ICAO and the Federal Communications Commission (FCC) in the United States. These organizations work together to ensure that adequate spectrum resources are available for aeronautical services and that they are used efficiently. In the United States, the FCC is responsible for managing spectrum allocation for aeronautical services. They work closely with the aviation industry to understand their spectrum needs and ensure that they have the necessary resources to operate safely and efficiently. The FCC also works with international organizations, such as the International Telecommunication Union (ITU), to coordinate spectrum allocation on a global scale. Figure 3.3 provides an overview of the spectrum

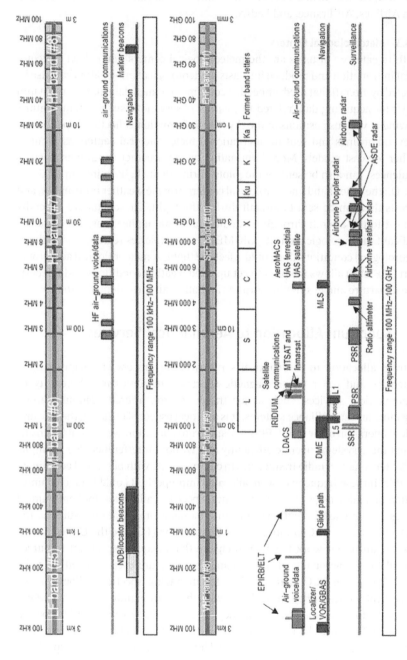

Figure 3.3 Overview of spectrum allocation to aeronautical services. Source: ICAO [2021]/International Civil Aviation Organization.

allocation process conducted by ICAO for aeronautical services. The allocation of spectrum to aeronautical services is a complex process that involves balancing the needs of different users and ensuring that there is no interference between different systems. To accomplish this, regulators use a variety of tools, such as spectrum sharing, frequency coordination, and interference mitigation techniques. As the aviation industry continues to grow and become more complex, the demand for spectrum resources will only increase. Proper spectrum allocation will be essential for ensuring the safety and efficiency of air travel in the years to come. Therefore, it is essential that regulators continue to work with the aviation industry to understand their needs and allocate spectrum resources effectively and efficiently.

In Sections 3.5 and 3.6, we will review the current airplane communication based on a study by Wichgers [2015] in 2015 to identify and evaluate communication technology candidates for NextGen National Airspace System (NAS).

3.5 Air-to-Air Communications

Existing NAS air-to-air communications are specifically limited to three main communication systems, traffic alert and collision avoidance system (TCAS), VHF Communications on the common traffic advisory frequency (CTAF), and automatic dependent surveillance-broadcast (ADS-B). In the following, we will discuss every section.

3.5.1 TCAS Communications

The TCAS uses two radio frequency carriers, 1030 and 1090 MHz, for its communication between suitably equipped aircraft. The system relies on the use of Mode S and Mode C transponders to transmit and receive signals. When a TCAS-equipped aircraft sends an interrogation request, it is transmitted on the 1030 MHz radio frequency carrier. The replies to these requests, known as "squitters," are transmitted on the 1090 MHz frequency carrier. Each Mode S transponder on a TCAS-equipped aircraft emits a unique squitter that pseudo-randomly radiates its Mode S address in all directions to let other like-equipped aircraft know of its presence. When a second aircraft receives the squitter, the TCAS system on that aircraft sends a Mode S reply back to the specific Mode S address contained in the squitter message. The use of directional antennas allows the TCAS system to determine the bearing to the neighboring aircraft. In addition to the Mode S transmissions, Mode C altitude broadcasts are used to establish the altitude of the nearby aircraft. The timing of the Mode S interrogation/response protocol is then used to determine the distance between the TCAS-equipped aircraft.

The bidirectional data link established between each TCAS-equipped aircraft is critical to obtain traffic information to support traffic situational awareness and traffic advisories for TCAS-I-equipped aircraft. For TCAS-II-equipped aircraft, the system can provide traffic resolutions to avoid collisions.

In summary, TCAS communication relies on Mode S and Mode C transponders to establish a bi-directional data link between TCAS-equipped aircraft. The use of unique squitters, directional antennas, and altitude broadcasts allows the TCAS system to determine the bearing, distance, and altitude of nearby aircraft, providing essential traffic information and advisories to ensure safe operations in crowded airspace.

3.5.2 VHF Communications

VHF voice communications are an essential component of aircraft operations at non-towered airports. These communications are used aircraft-to-aircraft and are carried out on a CTAF. At non-towered airports, pilots are required to broadcast their intentions and listen for other aircraft in the area on the designated CTAF frequency. This frequency is typically published in the airport directory or other appropriate documents. When approaching an airport, pilots will listen to the CTAF for any ongoing communications or announcements. Once they are within a certain distance from the airport, pilots will make a radio call announcing their position and intentions. This helps other aircraft in the area to be aware of their presence and intended actions, thus promoting safety in the airspace. Other pilots in the vicinity will also make similar broadcasts, allowing all aircraft to be aware of each other's movements and to adjust their flight paths accordingly. This helps to prevent collisions and promote safe operations in crowded airspace, where there is no ATC tower to provide direction. In summary, VHF voice communications on a CTAF are a crucial element of aircraft operations at non-towered airports. These communications allow pilots to announce their intentions and listen for other aircraft in the area, promoting safety in the airspace by preventing collisions and ensuring that all aircraft are aware of each other's movements.

3.5.3 ADS-B Air-to-Air Communications

ADS-B is an advanced communication system that allows aircraft and airport ground vehicles to exchange surveillance information using air-to-air (A-A) and air-to-ground (A-G) communications. ADS-B operates on two frequencies: 1090 MHz (as 1090 Extended Squitter – 1090ES) and 978 MHz (universal access transceiver). ADS-B is a data broadcast system that continuously transmits information about the aircraft's position, altitude, speed, and other relevant data to other aircraft and air traffic controllers. This information is transmitted using a Mode S transponder on 1090 MHz or a universal access transceiver on 978 MHz. ADS-B enables suitably equipped aircraft and airport ground vehicles to be tracked by pilots of other aircraft equipped with ADS-B receive equipment in a process called air-to-air (A-A) communication. This feature enhances safety by enabling pilots to see other aircraft on their cockpit displays, even when they are not in visual range.

3.6 Air-to-Ground Communications

Airborne communication systems that facilitate communication between aircraft and ATS, as well as AOC services, are known as air-to-ground (A-G) communications. These communication systems also support navigation, surveillance, and information services. The current A-G communication systems in the NAS include:

1. HF, VHF, and Satellite for supporting "communications" functions,
2. VHF data broadcast (VDB) for supporting "navigation" functions, and
3. L-band 978, 1030, and 1090 MHz.

Nowadays, it is mandatory for pilots to have direct voice communications with ATC in all airspaces. Within domestic airspace, which includes surface, terminal area, and domestic en route airspaces, VHF voice is the necessary ATS communication requirement. Some countries also offer ultra high frequency (UHF) voice to support military operations. On the other hand, in remote oceanic or continental polar airspaces that are beyond the reach of VHF voice communications, HF and satellites are utilized for communications. In the event of communication failure, flight crews follow the lost communication procedures and proceed with their flight plans. Further details on each of the previously mentioned A-G communications can be found in Sections 3.6.1, 3.6.2, 3.6.3, and 3.6.4.

3.6.1 HF Air-to-Ground Communications

HF voice communication is a crucial component of ATC for transoceanic flights and flights into remote areas where VHF A-G communication coverage is unavailable. HF radios enable voice and data communication over long-distance routes, including over the ocean and trans-polar regions. The use of HF communication acts as a safety network for beyond-line-of-sight long-distance data communication, complementing existing VHF and SATCOM. Unlike aircraft VHF communication, the spectrum for HF aviation communication is not divided into a large number of contiguous channels. The aviation allocation in the HF band is interspersed with many other services, so different frequencies must be used at different times of day and night and for different paths due to the variations in the height and intensities of the ionized regions of the earth's atmosphere. There is also seasonal variation, particularly between winter and summer, and HF propagation may be disturbed or enhanced during periods of intense solar activity. The frequencies selected for a particular radio path are usually set roughly mid-way between the lowest usable frequency (LUF) and the maximum usable frequency (MUF). Propagation at HF has considerable variation and is far less predictable than VHF propagation. The typical range for HF communication is 500–2500 km, effectively filling in the gap in VHF coverage. The available spectrum for aircraft communication at HF is extremely limited, so bandwidth is restricted to around 3.5 kHz for both voice and data transmission. For voice transmission, single sideband (SSB) modulation is used. HF data link (HFDL) uses M-ary phase shift keying (M-PSK) at data rates of 300 or 600 (for $M = 2$), 1200 (for $M = 4$), and 1800 ($M = 8$) bps per channel.

The HF data link is based on frequency division multiplexing (FDM) for access to ground station frequencies and time division multiplexing (TDM) within individual communication channels. Each Time division multiple access (TDMA) frame is 32 seconds and is divided into 13 equal slot durations. The first slot of each TDMA frame is reserved for the HFDL ground station subsystem to broadcast link management data. The remaining slots are designated as uplink slots, downlink slots reserved for specific HFDL aircraft, or downlink random access slots for use by HFDL aircraft on a contention basis. These TDMA slots are assigned dynamically using a combination of reservation, polling, and random access assignments. HF operates at SSB carrier frequencies available to the aeronautical mobile (R) service in the band of approximately 2–30 MHz. The following frequencies range in the HF band are allocated to aeronautical services, and they must be used by aircraft for ATC and communication:

1. 2850–3155 kHz
2. 3400–3500 kHz
3. 4650–4750 kHz
4. 5480–5730 kHz
5. 6525–6765 kHz
6. 8815–9040 kHz
7. 10,005–10,100 kHz
8. 11,175–11,400 kHz
9. 13,200–13,360 kHz
10. 15,010–15,100 kHz
11. 17,900–18,030 kHz
12. 21,870–22,000 kHz
13. 23,200–23,350 kHz

VHF A-G Communications will be discussed in greater depth in Chapter 5 due to the fact that we will be using one of the VHF data links (VDL mode 4) in order to transmit FDR/CVR recorder data.

3.6.2 Satellite Communications (SATCOM)

SATCOM systems play a vital role in aviation communication by relaying voice and data between aircraft and ground stations. These systems employ various types of satellites, including geostationary and low-earth polar orbits, depending on the service network. SATCOM is especially important for trans-oceanic and transpolar flights, where it supports a range of safety services. For instance, SATCOM facilitates voice communication between pilots and ATC, aircraft position reporting for air traffic control (ADS-C), and medical service communications that allow flight crews and passengers to speak directly with emergency room physicians at the medical service provider's response center.

SATCOM also supports advisory services such as weather and traffic (Sirius XM Satellite Radio), in addition to various data services. The Aeronautical mobile satellite services spectrum in the L-band supports ATS and AOC communications. Currently, the spectrum allocated includes parts of the L-band, ranging from 1525 to 1660.5 MHz. Various government and commercial SATCOM service providers exist, including major providers such as Iridium and Inmarsat. In Chapter 4, we will explain Iridium in depth, since we used its system as a case study to transmit data from FDR/CVR recorders. Inmarsat is an international satellite service provider that offers a range of voice and data services, including Aero H, H+, I, M, Swift 64, and the popular SwiftBroadband high-speed data service. Swift 64 and SwiftBroadband systems are commonly used in air transport and larger business aircraft due to the size of the high-gain antenna system typically installed on the tail of the aircraft. SwiftBroadband is an IP-based packet-switched service that provides "always-on" data at speeds of up to 432 kbps per channel, available globally, except for polar regions above 70° North/South latitude. It can also provide IP streaming at various rates up to a full channel. Inmarsat's satellite-to-subscriber frequencies range from 1525 to 1559 MHz and from 1626.5 to 1660.5 MHz. Equipping an aircraft with Inmarsat SATCOM enables a wide range of uses in the cockpit and the cabin, including aviation safety communications services, weather and flight-plan updates, passenger connectivity for email, internet access, voice over IP (VoIP) telephones, and Global System for Mobile Communication (GSM) and Short Message Service (SMS) messaging. Up to four channels per aircraft can be used. Inmarsat's Aviation safety services include flight deck safety (FDS) and enhanced group call (EGC) services, which provide critical information to pilots in real-time. The FDS service includes ADS-C and controller pilot data link communications (CPDLC), allowing for real-time data sharing between the aircraft and ATC. The EGC service is used for the broadcast of critical messages to multiple aircraft simultaneously, including Notices to Airmen, weather updates, and other safety-related information. Inmarsat's Aviation Safety Services include the following:

- *Classic Aero services and SwiftBroadband*:
  - Classic Aero services are accessible over both the Inmarsat-3 (I-3) and Inmarsat-4 (I-4) satellite systems.
  - SwiftBroadband services are provided with the I-4 satellites.
  - Global coverage ≤ 70° North/South latitude.
- *Classic services – Aero H*: Aero H provides packet data rates of up to 10.5 kbps for ACARS, FANS, and Aeronautical Telecommunication Network (ATN) communications and up to 9.6 kbps per channel for multichannel voice, fax, and data links through a high gain-antenna – anywhere in the global beams of the I-3 satellites. In addition to safety applications, other applications include passenger, operational, and administrative communications.
- *Classic services – Aero H+*: Offers all the features of Aero H, but uses the I-3 regional spot beams and 4.8 kbps voice codecs to deliver voice services at a lower cost. Outside of regional spot beams, Aero H+ terminals operate in the global beams in the same way as standard Aero H systems. Aero H+ is also available in the full I-4 satellite footprint.
- *Classic services – Aero I*: Use intermediate-gain antennas and the I-3 regional beams, providing multi-channel voice and 4.8 kbps circuit-switched data services. Aero I packet data is also available in the full I-4 footprint.

Inmarsat has deployed a global wireless broadband network called Inmarsat Global Xpress, which uses three Inmarsat-5 (I-5) satellites to provide full global coverage. The satellites operate at the Ka-band frequency range of 20–30 GHz and each carries a payload of 89 small Ka-band beams that offer global spot coverage. The network is designed to provide high-speed inflight broadband on airliners, among other applications. The deployment of the three satellites was completed by the end of 2014, as planned.

3.6.3   VHF Data Broadcast (VDB) Communications

The VDB is a data communication technology used to support high-accuracy Global Navigation Satellite System (GNSS) applications such as Category I/II/III precision approach operations. It is transmitted from the ground based augmentation system (GBAS), which is also known as the local area augmentation system (LAAS). The VDB broadcast includes GNSS differential correction information, satellite integrity data, final approach segment definition data, and ground station location data. The VDB link layer is similar but different from VDL-M2, and it complies with the physical layer of the ISO stack protocol described in ICAO Document AMCP/3-R/8A (VHF Digital Link Manual). The VDB uses the D8PSK modulation to achieve a 31.5 kbps nominal signaling rate, similar to VDL-M2. VDB is an emerging data communication in the NAS, with two operationally commissioned GBAS systems located at Newark Airport in New Jersey and Houston Airport in Texas. These systems were commissioned by the FAA, and the first one was approved for operational service on 28 September 2012. In addition to these operational systems, there are several prototype and test systems in development. The air traffic control radar beacon system (ATCRBS) uses several modes for aircraft identification and altitude reporting. Mode 1 is used to sort military targets, while Mode 2 identifies military aircraft missions. Mode 3/A identifies each aircraft in the radar's coverage area, and Mode C is used to request/report aircraft altitude. Modes 4 and S use the same transponder hardware but are not considered part of the ATCRBS system. Mode 4 is used by military aircraft for the Identification Friend or Foe system, while Mode S is used for discrete selective interrogation to facilitate TCAS for civilian aircraft. Mode S differs from ATCRBS in that each aircraft is assigned a unique address code, allowing interrogations to be directed to a specific aircraft and replies to be unambiguously identified. The Mode S interrogator provides surveillance of all beacon-equipped aircraft within its line of sight using binary differential phase shift keying (DPSK) and consists of a 24-bit discrete address. Mode S can provide air-to-ground and air-to-air data links and is primarily used for surveillance. The TCAS-I maximum range is 200 NM. The Mode S transponder communicates using short (56-bit) or long (112-bit) extended squitters, and longer messages can be transmitted using the extended-length message (ELM) capability. ADS-B/ADS-R/TIS-B (A-G and A-A), TCAS (A-A), and other ATC use for air and surface surveillance all use extended squitters.

3.6.4 ADS-B/ADS-R/TIS-B Air-to-Ground Communications

ADS-B is a system that broadcasts an aircraft's surveillance information to other aircraft and ground-based receivers, transmitted on two different frequencies, 1090 MHz (1090 Extended Squitter – 1090 ES) and 978 megahertz (Universal Access Transceiver – UAT). ADS-B is complemented by two other emerging ground-based traffic information services: ADS-B and Traffic Information Service – Broadcast. Automatic Dependent Surveillance – Rebroadcast relays ADS-B information transmitted by an aircraft using one link technology to another using a different technology. Traffic Information Service – Broadcast broadcasts traffic surveillance information for aircraft/vehicles that are not broadcasting ADS-B surveillance information. All three systems use pulse position modulation encoding of message data, with a pulse transmitted in the first half of the interval representing a ONE and a pulse transmitted in the second half representing a ZERO. For further information on transmission, readers can refer to the latest versions of RTCA/DO-260 for 1090 ES and RTCA/DO-282 for UAT. Another system that has emerged is the flight information service-broadcast, which provides pilots with weather and flight information, including graphical and textual weather data, as well as Notices to Airmen. The Flight Information System Broadcast (FIS-B) service is provided using the UAT data link, which is also used for implementing ADS-B OUT in the United States. FIS-B ground stations receive weather and aeronautical data from various sources and generate sets of products specific to their location and region of interest, broadcast to equipped aircraft free of charge. FIS-B provides pilots with critical information for flight planning and decision-making, improving situational awareness during flights. Current FIS-B products include Airmen's Meteorological Information, Significant Meteorological Information, Convective SIGMET, Meteorological Aviation Routine Weather Report, Continental United States Next-Generation Radar, Regional NEXRAD, Notice to Airmen, Pilot Report, Special Use Airspace Status, Terminal Aerodrome Forecast, Winds and Temperatures Aloft, and Traffic Information Service Broadcast (TIS-B) Service Status.

3.7 Summary

In this chapter, we explored the history and development of aviation communication systems, tracing their evolution from the earliest days of aviation to the present day. Drawing on insights from Mahmoud et al. [2014], we provided an in-depth analysis of the different types of communication traffic in the context of aircraft communication networks, including voice, data, and surveillance. We also highlighted the key players and organizations involved in this field and their critical role in ensuring effective communication between pilots, air traffic controllers, and other stakeholders.

To provide a more detailed understanding of the current state of aviation communication, we relied mainly on a study conducted by Wichgers [2015]. This study provided a comprehensive overview of the industry's challenges and opportunities, including the need for greater interoperability and standardization across different systems and technologies. By synthesizing this information with our existing knowledge, we were able to offer a comprehensive and up-to-date analysis of the state of aviation communication.

References

Editorial Team. The History of Airline Industry, 9 August 2023. URL https://www
.allgetaways.com/flight-booking/history-airline-industry.html.

Henri Farman, 2023. URL https://www.britannica.com/biography/Henri-Farman.

Flying the Beams. Early radio tech, 2020. URL https://flyingthebeams.com/early-
radio-tech.

W. Higginbotham. The birth of powered flight and air-to-ground communications,
April 2018. URL https://www.afspc.af.mil/News/Article-Display/Article/1514245/
the-birth-of-powered-flight-and-air-to-ground-communications/.

ICAO. Handbook on radio frequency spectrum requirements for civil aviation,
2021. URL https://www.icao.int/safety/FSMP/Documents/Doc9718/Doc.9718
%20Vol.%20I%20(AdvanceUneditedVersion%202021).pdf.

M. Mahmoud, C. Guerber, N. Larrieu, A. Pirovano, and J. Radzik. *Aeronautical
Air–Ground Data Link Communications*. John Wiley & Sons, Inc., 2014.

J.M. Wichgers. *Identification and Analysis of Future Aeronautical Communications
Candidates*. NASA technical memorandum. National Aeronautics and Space
Administration, Glenn Research Center, 2015. URL https://books.google.com/
books?id=4rV5zgEACAAJ.

4

Satellite Data Transfer Implementation

In this chapter, considering the Iridium system capabilities, we analyze the capability of the first-generation and second-generation Iridium satellite systems to carry the entire aircraft data load under different implementation scenarios. In each of the satellite implementation schemes, the aircraft in the system will attempt to transfer data to the Earth via satellite links. Every plane will transmit its data directly to the satellite without any relaying or forwarding via any element in the model. Through this analysis, we aim to determine the methods that can be used to maximize the use of the available spectrum and accommodate as many planes as possible while utilizing as little spectrum as possible. Toward the end of the chapter, we will explore the use of emerging Low Earth Orbit (LEO) satellite networks to transmit critical aircraft data to a ground base station, providing a potential solution to the current limitations of Iridium approaches.

4.1   The Iridium Satellite System

Iridium is a global provider of mobile satellite services. The network consists of a resilient LEO satellite constellation of 66 satellites plus spares in orbit to serve in case of failure. The Iridium satellite system was conceived and designed by Bary Bertiger, Dr. Raymond Leopold, and Keneth Peterson of Motorola [Nelson, 1998]. Iridium was originally designed to carry voice data. The system's backbone consists of four major components: the constellation of LEO satellites, Iridium Earth stations (Gateways), Iridium system control facilities, and Iridium subscriber units. Iridium was supposed to be made up of 77 satellites linked together worldwide, much like the 77 electrons orbiting around the nucleus of the element Iridium, hence the name. Later, the designers decided that the current set of 66 satellites was sufficient to provide global coverage. The first generation of 66 Iridium satellites, along with six spares, was launched between May 5, 1997 and May 17, 1998 [Nelson, 1998]. Iridium was intended to be used as

Real-Time Ground-Based Flight Data and Cockpit Voice Recorder: Implementation Scenarios and Feasibility Analysis, First Edition. Mustafa M. Matalgah and Mohammed Ali Alqodah.
© 2024 The Institute of Electrical and Electronics Engineers, Inc. Published 2024 by John Wiley & Sons, Inc.

an alternative to existing terrestrial cellular phone systems, with the advantage of having worldwide coverage [Matale, 2010]. The Iridium constellation was replenished with the Iridium NEXT system. The satellites have been launched in eight stages, starting in January 2017 and finishing in January 2019, and have successfully replaced the previous generation of Iridium satellites. Iridium NEXT provides a complete global satellite communications coverage service and offers more subscriber capacity, faster data speeds, and the ability to host payloads [eoPortal, 2013].

4.2 Iridium First Generation

The first generation of constellation technology was designed by Iridium Communications Inc. (formerly known as Iridium Satellite LLC) and sponsored by Motorola. The satellites' orbital missions began in 1997. Before all of the satellites were in orbit, commercial service couldn't begin. Iridium has employed Launch vehicles from various countries to put satellites into orbit, including the United States, Russia, and China. The first-generation fleet was estimated to have cost over US $5 billion [Graham, 2018]. The first test call over the network was made in 1998, and by 2002, the whole world was covered. In terms of technology, the system achieved its goals but was not generally adopted by the public. There was poor reception indoors, the handsets were inconvenient and pricey, and ordinary cellular phones were the rule in the market at the time [McIntosh, 1999]. As a consequence, Iridium went bankrupt, making it one of the largest bankruptcy in US history since its income wasn't enough to cover the costs of developing its constellation [Millard, 2016, Graham, 2018].

Once the original Iridium company went bankrupt, the constellation kept on. A new entity evolved to operate the satellites and established a different product placement and pricing strategy, providing communication services to a niche market of clients who wanted this reliable service in areas of the Earth not served by regular geosynchronous orbit communication satellite services. Journalists, explorers, and the military are just some of the users [Millard, 2016]. Even though the original satellites were expected to have a short design life of only eight years, no new satellites were launched between 2002 and 2017 to replace the constellation.

In the following, we describe the technical specifications, channels, and data rates of the Iridium system.

4.2.1 Technical Description

Iridium's satellites, weighing approximately 700 kg each, revolve around the Earth in six near-polar orbits of about 780 km above the surface of the Earth

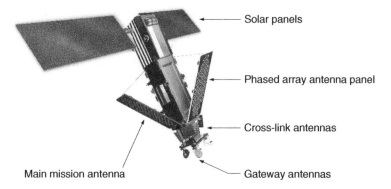

Solar panels

Phased array antenna panel

Cross-link antennas

Main mission antenna

Gateway antennas

Figure 4.1 Iridium satellite Matale [2010]. Credit: Lockheed Martin Corporation.

[Matale, 2010]. With 66 in-service satellites, each orbital plane contains 11 satellites. The first and last planes are spaced 20° apart and rotate in opposite directions creating a "seam," and the other co-rotating planes are spaced 31.6° apart. Each of the satellites, as shown in Figure 4.1, has a main mission antenna panel, a gateway antenna, and cross-link antennas [Lemme et al., 1999].

The main mission antenna consists of three phased array antennas with receive/transmit modules. The main mission antenna communicates directly with mobile units through tightly focused antenna beams. Main mission antennas operate in the L-Band of frequencies between 1616 and 1626.5 MHz. The satellites use the gateway antennas to communicate with one of the many gateways on Earth, eventually linking satellites with existing terrestrial Public Switched Telephone Network (PSTN). These Gateway antenna use the K-Band frequencies between 19.4 and 19.6 GHz as well as the band between 29.1 and 29.3 GHz. The cross-link antennas are used for inter-satellite networking. In sending data around the globe, Iridium satellites can route data from satellite to satellite, until a satellite can connect to the most appropriate gateway, based on the destination of the data [Pratt et al., 1999]. Table 4.1 shows the available frequency bands.

With the main mission antennas, the gateway antennas, and the cross-link antennas, each of Iridium's satellites has the ability to communicate directly with

Table 4.1 Satellite communication frequency bands.

L-Band up/down-link (satellite to/from mobile)	1616–1626.5 MHz
K-Band downlink (satellite to gateway)	19.4–19.6 GHz
K-Band up-link (gateway to satellite)	29.1–29.3 GHz
Ka-band Cross-link (satellite to satellite)	23.18–23.38 GHz
La-Band (satellite to/from system control)	1452.96 MHz

a mobile unit, directly with one of the Gateways, or relay data to the satellites immediately around it.

Each satellite in the Iridium constellation has up to four cross-links or inter-satellite links [Fossa et al., 1998]. One transmits data to the satellite immediately ahead of it while another transmits to the satellite directly behind it, both on the same orbital plane. The other two are cross-plane links, transmitting data to satellites on either side of them on different planes.

There are two main designs for satellite networks. In one of the designs that is used by satellite systems like Globalstar, each of the satellites acts like a mirror and is used simply as a radio repeater. In this design, the repeated signal can only be bounced back to the same location covered by the satellite beam where the signal originated. This means that for a successful call to be completed, a gateway satellite must be available for every coverage beam. The design used by the Iridium system employs intelligent satellites with complex routing and switching algorithms to support satellite interlinking. This inter-satellite linking capability affords Iridium the ability to route data from satellite to satellite until it can be downloaded to an appropriate gateway or mobile unit anywhere on the globe [Jabbar, 2001].

The system control for Iridium shown in Figure 4.2 is responsible for monitoring, managing, and controlling the satellites in the constellation. Iridium's

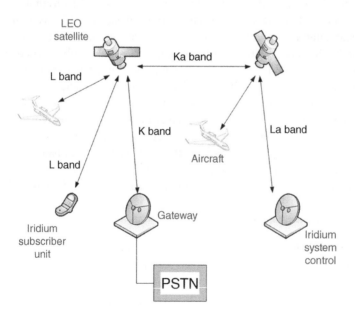

Figure 4.2 Iridium system overview.

Satellite and Networks Operations Center is located in Lansdowne, Virginia, United States of America, while the Tracking, Telemetry and Control (TTC) stations are located in Canada, Iceland, and Hawaii [Lemme et al., 1999]. The Iridium Satellite Network Operation Center (SNOC) is responsible for developing and distributing the routing tables that the satellites use to route data efficiently.

Gateways provide all call processing and the interconnections between the satellite system and the PSTN [Lemme et al., 1999]. The gateways act much like the Mobile Switching Center (MSC) in cellular networks, also providing services like subscriber validation and call processing.

4.2.2 Channels

Communication between aircraft and the satellite will be facilitated by the satellite's main mission antennas, with each aircraft being a mobile station.

The main mission antennas on the satellites that are responsible for communicating directly with mobile devices are made up of three panels of phased array antennas. The first panel is perpendicular to the travel of the satellite, the second is displaced 120 counterclockwise from the first panel, while the third is displaced 120 counterclockwise from the second. Each of these panels has the ability to transmit 16 tightly focused spot beams, giving each satellite a total of 48 spot beams (16 beams per panel × 3 panels = 48 beams) [Hubbel and Sanders, 1997]. With 66 satellites in the constellation, the entire system can produce up to (48 beams per satellite × 66 satellites) 3168 beams. Not all of these beams are used by the system. It so happens that because of the overlap of beam coverage that occurs as the satellites move to higher latitudes from the equator, the system only needs to use 2150 beams at a time to provide global coverage [Hubbel and Sanders, 1997]. The switching on and off of the beams as the satellites move to and from the poles is taken care of by complex routines carried out by the SNOC.

Just as in the case of cellular coverage, beams can be grouped into clusters. Each cluster is assigned a set of channels or frequencies. Different clusters can later reuse the same set of channels. In the Iridium, 12 beams are used to form a cluster, like a 12-cell cluster in cellular terminology [Hubbel and Sanders, 1997].

The Iridium system utilizes 10.5 MHz of bandwidth in the L-Band spectrum of frequencies (Figure 4.3). This spectrum is divided into 240 FDMA channels each with a bandwidth of 31.5 and 10.17 kHz of guard band to minimize intermodulation effects and leakage. Channels are placed 41.67 kHz apart. Additional guard bands of 37.5 kHz are used to allow for Doppler shifts.

Within each of these 240 FDMA channels, there are 8 Time Division Multiple Access (TDMA) time slots of duration 8.28 ms each, inside a 90 ms frame as shown in Figure 4.4. When iridium is used for two way voice communication, four of

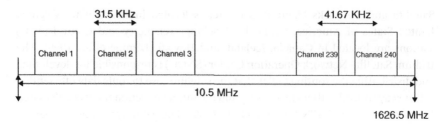

Figure 4.3 FDMA scheme for Iridium.

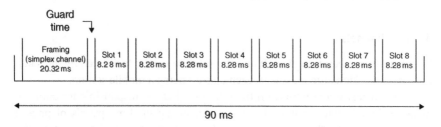

Figure 4.4 TDMA scheme for Iridium.

these times slots are designated uplink, and the other four are used for downlink communication.

A pair of uplink/downlink time slots constitutes one Full Duplex (FDX) channel. There are, therefore, 4 FDX channels per FDMA channel. With 240 FDMA channels, the total number of full duplex channels available is $240 \times 4 = 960$. Each cluster of 12 beams can replicate these 960 channels, giving each beam an 880-channel load capacity [Hubbel and Sanders, 1997]. However, the Iridium system only uses 65 of these channels as voice channels, leaving 15 FDX channels (or 30 simplex channels), equaling 18.75% of the spectrum, for data per beam. The Iridium system is capable of routing data around the globe through its satellite cross-links. With this in mind, we assume that the slots in each frame that we analyze are purely uplink slots because the satellites do not reply to every transmission from the aircraft back to the aircraft, rather down to one of the gateways, using another band of frequencies with independent channels. Therefore, $30 \times 12 = 360$ simplex channels are available to be used as one-way data channels per cluster of 12 beams, and because 2150 Iridium beams are active in the system at any given time, the number of channels available for data transfer is 64,500 over the entire system ($2150 \times 30 = 64,500$).

4.2.3 Channel Data Rate

The Iridium system employs the Quadrature Phase Shift Keying (QPSK) modulation scheme. Ideally, QPSK modulation allows two bits of data to be represented

Figure 4.5 Quadrature phase shift keying constellation diagram.

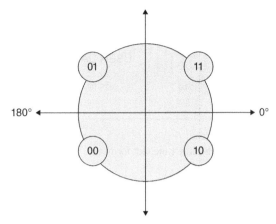

as one signal. That is, two bits of data can be transmitted in one signaling interval. Data is represented in QPSK by modifying the phase of the carrier signal, with four different possible values. A constellation diagram for QPSK modulation is shown below.

At each signaling interval, a theoretical 2 bits of data can be transmitted when using QPSK (Figure 4.5). Ideally, QPSK modulation achieves the transmission of 2 bits per signaling interval (or per hertz of spectrum). With a channel bandwidth of 31.5 kHz, QPSK modulation should be able to deliver 63 kbps. However, because of the division of the channel into time slots, and the use of some of the time for guard time between slots, the bit rate is slightly reduced. With eight time slots, each with a length of 8.28 ms, used for data transfer, the channel only delivers data for a percentage of the time. The achievable data rate of the QPSK modulation in the satellite communication band of frequencies, in conjunction with the use of raised cosine filtering, is 50 kbps [Nelson, 1998]. Each 90 ms frame has about 71.42 ms that is used for transmission, while the rest of the time is used for framing and guard slots. Having a bandwidth of 31.5 kHz, each of the FDMA channels is able to achieve a bit rate of 50 kbps ((71.42 ms/90 ms) × 2 bpHz × 31,500 Hz = 50,000 bps). The data rate for each of the time slots of length 8.28 ms inside a 90 ms frame is theoretically equal to (8.28 ms/90 ms) × (50 kbps) = 4.588 kbps.

This is a raw data rate and does not quantify the amount of data that can be transferred or transmitted through the channel. As with any data communication system, the Iridium system requires each of the bursts or time slots to have overhead bits, including source address/destination address, and other signaling messages to be included in each transmission. A Forward Error Correction (FEC) code of rate (3/4) [Nelson, 1998] is included in the system's design. This means that, exclusive of code bits, we can transmit information at a rate of (0.75 × (8.28 ms/90 ms) × (50 kbps) = 3.45 kbps). The eight time slots used for

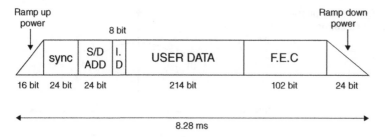

Figure 4.6 Iridium time slot format.

data transmission are housed in a 90 s time frame. 8.28 ms × 8 time slots = 66.24 ms. This leaves 23.76 ms (90 ms − 66.24 ms = 23.76 ms) of the TDMA frame for framing bits and guard time slots, as well as also being configured to be used as a Simplex channel that can be used by the satellite to communicate with mobiles. Therefore, this excess time in the frame can be used to send acknowledgments (ACKs) and resend requests (REQs) from the satellites. Specifics about the configuration of Iridium's time slots are not published in the open literature, and some assumption, and generalizations are made to attain a good estimate for the throughput of the system. In this book, we assume that each time slot will include 16 bits of ramp-up power, 24 bits of synchronization, 24-bit source/destination address, a 24-bit ramp-down power sequence, and 8 bits of message ID at least, in addition to the FEC bits added. We also work with the assumption that only 11 frames are available inside of each second.

With each frame being 90 ms long, and with each user transmitting once in every frame, we can see in Figure 4.6 that each time slot can transmit a net bit rate of 4588 bits/(1/90 ms) = 412 bits per time slot, or 3450 bits/(1/90ms) = 310 bits per time slot excluding FEC, guard bits, and power ramp down bits. This implies that, using our assumptions for the configuration of Iridium system bursts, about 102 bits are used for error checking and correction for each of Iridium's transmissions.

When we subtract the 96 bits of overhead and signaling per burst, we have an information bit rate of 214 bits per time slot, resulting in a rate of 214 bits per timeslot × 1/90 ms ≈ 2354 bps.

4.3 Second Generation

In February 2010, Iridium awarded Thales Alenia a US $2.3 billion contract to design and manufacture 81 satellites for the Iridium NEXT project. In the same year, Iridium and Space Exploration Technologies (SpaceX) inked a US $492M contract, making SpaceX a major launch provider for the NEXT mission.

Figure 4.7 Iridium NEXT satellite. Source: [iridium, n.d.].

Iridium intended to employ SpaceX's Falcon 9 launch vehicle to place a fleet of second-generation satellites into LEO, as part of the company's plan to upgrade its first-generation satellite constellation. Figure 4.7 illustrates one of the second-generation Iridium satellites.

In January 2017, the first Iridium NEXT satellite was launched, and in February of the same year, the first satellite of the next generation started operating. As of the eighth and final launch of the Iridium NEXT campaign in February 2019, all 10 of the second-generation satellites had been placed into a geostationary transfer orbit. As part of the Iridium NEXT program, 75 second-generation satellites were sent into space, and 66 are now part of the operational constellation. The final satellite of the second generation entered service on 5 February 2019. These satellites, once launched and fully operational, marked the completion of the largest space technological upgrade in history [iridium, n.d.].

Although the Iridium satellites are primarily intended to support the Iridium communications mission, they have been modified to handle hosted payload missions. The K-band network of cross-links between satellites, feeder links to the ground, and teleports connecting the satellites through Earth stations to an MPLS (Multiprotocol Label Switching) cloud called the Teleport Network can

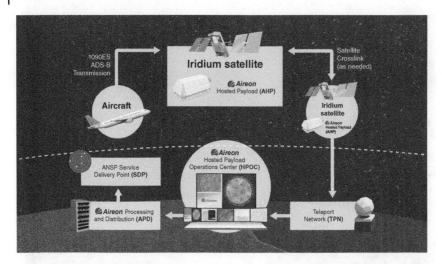

Figure 4.8 Space-based automatic dependent surveillance-broadcast (ADS-B) system. Source: Aireon [2018]/ https://aireon.com/resources/overview-materials/technical-specifications last accessed February 25, 2023.

transport mission data, sensor telemetry, and command data for these missions in near real-time [eoPortal, 2013].

It is important to point out that Iridium NEXT's partnership with Aireon Space's satellite constellation is notable since it is the only satellite constellation that provides the orbital configurations required to carry out worldwide air traffic surveillance successfully. Space-based Automatic Dependent Surveillance-Broadcast (ADS-B) in Figure 4.8 provides unparalleled global surveillance coverage to receive and process ADS-B signals broadcast from aircraft equipped with 1090 MHz ADS-B transponders. Without the need for ground stations, this setup can cover the whole planet, including the oceans and the polar regions. There is currently no other system in place or in the works that provides this possibility for the aviation industry. When an airplane transmits ADS-B data, the Harris-built Hosted Payload will receive it and relay it to the Aireon Teleport Network and Aireon Processing and Distribution (APD) system on the ground. The APD decodes and verifies the data with the help of its partner, Harris, and then sends it on to the appropriate stakeholder facilities that have subscribed for the Aireon service [Aireon, 2018].

4.3.1 Orbit

The Iridium-NEXT constellation is identical in architecture to the original system. As shown by the RF footprints simulation in Figure 4.9. The 66-satellite primary constellation, along with the six spares already in orbit, can provide continuous

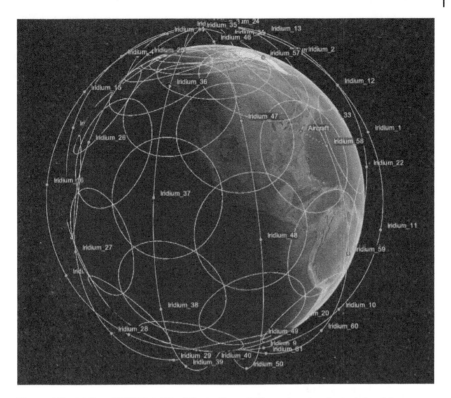

Figure 4.9 Iridium NEXT Satellite (Illustration of RF overlapping footprints of the Iridium NEXT satellite constellation). Source: [iridium, n.d.].

and reliable global coverage. The main constellation is organized into six orbital planes, each with 11 slots that are evenly spaced apart as shown in Figure 4.10. This is made possible by using cross-linked satellites that work together to form a fully meshed network. This network is maintained by many spare satellites in orbit, and it transmits real-time data downlinks to a ground station network operated by Iridium. The lifetime of the constellation is estimated to be greater than 10 years when it is placed in a polar orbit at 780 km in altitude with an inclination of 86.4° [eoPortal, 2013].

4.3.2 Spacecraft

The Iridium-NEXT satellites are based on Thales Alenia's ELiTeBus-1000 (Extended LifeTime Bus) satellite platform, each weighing 860 kg at launch and measuring 3.1 by 2.4 by 1.5 m in size in its stowed configuration. The Iridium-NEXT satellites have been designed for an operational life of 12.5 years

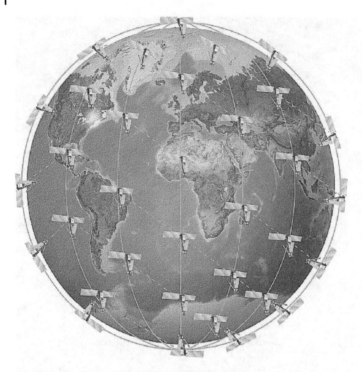

Figure 4.10 Iridium NEXT Satellite (Orbital coverage of the Iridium NEXT constellation of 66 spacecraft. Source: [iridium, n.d.].

with a stretch goal of 15 years, holding sufficient propellants to support an extended mission [Spacecraft & Satellites website, 2018]. Detailed specifications of Iridium NEXT spacecraft are listed in Table 4.2.

4.3.3 Characteristics and Communication Links

Iridium employs a communication system known as Differentially Encoded Quadrature Phase Shift Keying (DEQPSK), which utilizes a bandwidth of 31.5 kHz. To ensure proper reception of Iridium signals, each channel is spaced 41.667 kHz apart, which represents the minimum required bandwidth for receivers. Iridium employs a combination of TDMA and Frequency Division Multiple Access (FDMA) in its transmissions. The transmitter duty cycle of Iridium allows for burst transmissions occurring every 8.28 ms within a 90-ms time frame, which corresponds to a rate of 9.2%. These bursts are transmitted at a speed of 50 kilobits per second, with a symbol rate of 25,000 baud. For voice

Table 4.2 Specification of Iridium NEXT spacecraft [eoPortal, 2013].

Spacecraft launch mass, power	860 kg, 2 kW
Deployed wingspan	9.4 m
Spacecraft size (launch configuration)	3.1 m × 2.4 m × 1.5 m
Mission life	10 year design and 15 year mission life
Spacecraft stabilization	2-axis attitude control. A total of 248 AA-STR (Active Pixel Sensor-based Autonomous Star Tracker) are being supplied by Selex Galileo for the Iridium NEXT comsat constellation of 66 satellites
RF communications	– Regenerative processing payload with OBP (On-Board Processor) – Single 48-beam transmit/receive phased array antenna
L-band	– TDD (Time-Division Duplex) architecture
Ka-band	– Two 20/30 GHz steerable feeder links to terrestrial gateways
	– Four 23 GHz cross-links to adjacent Iridium NEXT satellites for relay communications (with two steerable, two fixed antennas, and TDD architecture)
TT&C	– 20/30 GHz links via omni antennas
Orbital altitude of constellation	780 km

communications, Iridium utilizes a specialized vocoder called Advanced Multi-Band Excitation (AMBE), developed by Digital Voice System Inc. This vocoder operates at a rate of 2.4 kilobits per second and is specifically designed to optimize voice transmission over the Iridium communication channel.

4.3.3.1 The Subscriber Links

The subscriber links use Time Division Duplex (TDD) between the uplink and downlink signals to provide mobile satellite service at 1616.0–1626.5 GHz. In the 1616–1626.5 MHz uplink and downlink bands, there are 252 carriers with carrier spacings of 41.667 kHz (240 primary carriers and one ring alert carrier at 1626.2708 MHz, four carriers for paging and acquisition, and seven unused carriers as guardbands). Each of these carriers requires a bandwidth of 35 or 36 kHz. These carriers may be combined to provide signals with a required bandwidth of up to 288 kHz. These 252 carrier frequencies can be grouped into eight carrier subbands. Each uplink carrier is independently assignable to each downlink carrier. Right Hand Circular (RHC) polarization is utilized in both the uplink and downlink 1616–1626.5 MHz bands [FCC, 2013].

4.3.3.2 The Feeder Links

The Ka-band Iridium NEXT feeder links operate at (19.4–19.6 GHz) for the downlink and (29.1–29.3 GHz) for the uplink as illustrated in Figure 4.11. There are 13 uplink and downlink transponders/channels with a required bandwidth of 14 MHz. It is possible to assign the feeder uplink transponders independently of the feeder downlink transponders. These feeder link transponders use RHC polarization in the uplink and Left Hand Circular (LHC) polarization in the downlink [FCC, 2013].

4.3.3.3 The Inter-Satellite Links

Each Iridium NEXT satellite is equipped with four inter-satellite links that allow it to communicate with its neighbors in the same orbit, as well as those in the surrounding orbits. In the frequency range from 22.18 to 22.38 GHz, eight separate transponders/channels are used for sending and receiving data across inter-satellite links. These transponders require a bandwidth of 21.6 MHz and have center frequencies separated by 25 MHz. A satellite's transmit and receive inter-satellite links can be designated separately from one another. Horizontal polarization is used for both the transmission and receiving of inter-satellite

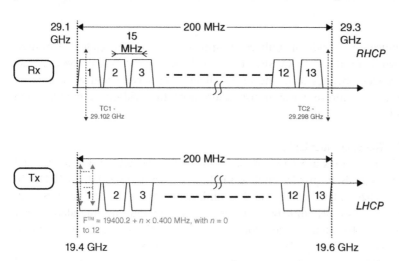

Figure 4.11 Iridium NEXT Ka-band frequency and polarization.
Source: [FCC, 2013]/Public Domain.

connection frequencies. These connections are only half-duplex since they use the same carrier frequency for transmission and receiving.

4.3.3.4 The Telemetry, Tracking, and Commanding (TT&C) Links

The TT&C links deliver data concerning the attitude, status, and functionality of the satellite. RHC polarization is utilized by both of the available telecommand carriers. These telecommand carriers operate on center frequencies 29,102 and 29,298 MHz and have a required bandwidth of 1 MHz. There are 13 telemetry carriers with required bandwidths of 200 kHz and channel spacings of 400 kHz that use LHC polarization and center frequencies ranging from 19400.2 to 19405.2 MHz.

The L-band mobile-satellite service (MSS) carriers, the Ka-band feeder links, and the Inter-satellite Service (ISS) links all use processors built into the Iridium NEXT satellites. Any of the 48 L-band beams can be given any of the 252 L-band carrier frequencies at any time. The five simplex carriers/channels in the 1626–26.5 GHz range can and will be dynamically assigned to each of the 48 L-band beams.

4.3.4 Band Frequency Reuse

In the L Band, the Iridium NEXT system supports two types of frequency reuse: At the constellation level, between satellite beams that are sufficiently separated to avoid unacceptable interference between the satellites, at the level of a single satellite, between two independently isolated beams.

Frequency reuse is determined by the traffic handled by the constellation or a portion of the constellation. The strategy is to evaluate all potential deployments of frequency resources and select the one with the lowest interference cost. On the same spacecraft, you can't give an L-band carrier to all 48 beams simultaneously. Each satellite can assign only 32 out of the 48 beams in any one time slot.

When in near-polar orbits, Iridium satellites are closer together. As satellites continue their orbital path toward the poles, their coverage areas overlap. By turning off individual satellites' outer ring spot beams at high latitudes, the workload is evenly distributed across the satellites. This will also reduce inter-satellite interference and minimize satellite power consumption while maintaining Earth coverage.

4.3.4.1 TDMA Frame Structure

Iridium NEXT uses the same TDMA Frame structure (Figure 4.12) as the first generation, in which a frame is 90 ms and time slots are assigned to simplex channels and duplex uplink and downlink carriers. The simplex time slot is 20,300 ms, while each of the four uplink and downlink time slots is 8267ms. Guard times are also included in the frame.

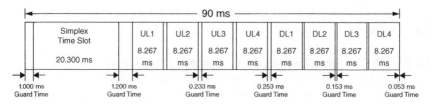

Figure 4.12 TDMA frame structure Source: [FCC, 2013] /Public Domain.

4.4 PSTN-Based Data Transfer Implementation: One Channel per Aircraft

Even though The Iridium Satellite system provides worldwide coverage, we would like to evaluate its data transfer capabilities in an area with heavy air traffic. The solutions shown for one particular region are not unique to that area. Results obtained from studying the capacity of the system in one area are applicable to any similar area anywhere on the globe. We shall provide a local and regional case study for studying capacity (United States), while these findings are considered relevant elsewhere in the world.

Each of Iridium's satellites has the ability to cover an area that is about 4400 km [Fossa et al., 1998] in diameter or 15,205,308 km^2. For prospective, the United States has an area of about 9,161,923 km^2 (or 3,537,441 m^2), and can therefore be covered by one satellite. However, Iridium beam coverage areas overlap each other slightly at the edges, and the North American continent and any other location with a similar area can be covered by up to four Iridium satellites.

The 240 frequency bands (FDMA channels) in the Iridium system can be reused every 12 beams, with every satellite providing 48 beams. The combination of these four satellites will therefore be able to provide ($240 \times 4 \times 4 = 3840$ channels). Data traffic uses 18.75% of the available channels, and channels for data transfer will therefore be able to use 720 FDMA channels (180 per satellite or 45 per cluster). Each of these channels is capable of supplying eight simplex data transmit slots, resulting in a total number of 5760 slots for data communication.

We will assess the transfer capability of the system provided in Figure 4.13 and determine how many aircraft can be accommodated using Iridium satellites. We assume that data channels from four satellites will be available for use in an area the size of the North American Continent. From the derivations given before, and knowing that each satellite produces 48 beams, we see that each satellite can offer 360 channels per 12 beams $\times$ 4 = 1440 available data channels because each of the 4, 12 beam clusters can provide 360 simplex data slots. With four satellites per observation area, the total number of channels available in the area

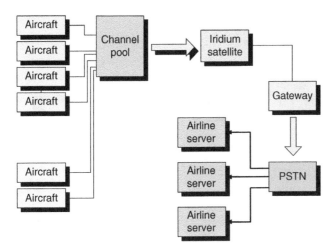

Figure 4.13 System model for PSTN-based implementation.

is $1440 \times 4 = 5760$. Each data channel can convey 2354 bps, meaning that four satellites with their combined 192 beams can carry a maximum load of 5760×2354 bps $= 13,559,040$ bps. For average air traffic conditions in the area, there are a sufficient number of channels available to allow each mobile (aircraft) to have a dedicated channel.

With each aircraft generating 1536 bits of data every second, and every available data channel offering a capacity to carry a load of 2354 bits per second, the first clear case is that in any case where traffic density is not extreme, the Iridium system is capable of handling the entire load of data, in real time by offering one data channel per aircraft. The maximum number of planes that can use this system, at one plane per time slot $= 13,559,040/2354 = 5760$ aircraft. This implementation supports 8 planes per 31.5 kHz channel, one for each of the 8 time slots per 90 ms frame. Access to these time slots would be assigned and dedicated for the duration that the aircraft uses the channel for streaming, with each plane transmitting once every frame.

The method of implementation assigns one transmit slot per aircraft in Figure 4.14 is obviously inefficient, because each transmit slot can carry more data than the aircraft actually needs to send. The extra 818 bits in the transmission are not required and are therefore wasted. What we would like to have is a system that allows aircraft to share these time slots in order to utilize the capacity to its maximum. Given that 13,559,040 bits can be sent through the system, and ensuring that all (as much of) this availability is used for aircraft data, the number of aircraft that can be supported can increase to a theoretical maximum of about 8827 ($13,559,040/1536 \approx 8827$).

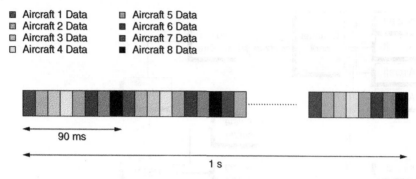

Figure 4.14 Single channel TDMA scheme for 1 plane per channel utility.

Finding the maximum number of planes that can operate given a fixed resource pool is an important goal of this work. In order to find the maximum capacity, a method of channel sharing will have to be assigned other than the one slot per aircraft implementation. One applicable method is to employ buffer and burst techniques. In this scheme, each aircraft may be assigned more than one slot per frame at staggered intervals.

4.5 Alternative Satellite Transmission Implementations

The obvious lack of efficiency of the dedicated channel assignment leads us to analyze different ways in which the channels available can be used to achieve greater resource efficiency. We begin by assessing a method of channel sharing that involves assigning a fixed number of slots for each aircraft that is sharing a channel to use.

4.5.1 Fixed Slot Allocation per Aircraft per Burst

The first alternative transmit method of channel utilization calls for not assigning each aircraft an entire channel every one second. This would entail dividing the slots on necessity-based criteria. Each time slot can carry 214 bits of user data. The minimum required number of time slots needed to transmit 1536 bits of user data is therefore eight slots in one second, although this number will be varied in order to find the most efficient assignment. The time required for each burst with a different number of slots per burst in this implementation is therefore variable.

This implementation method assigns a fixed number of slots for each aircraft to use in every burst. These bursts do not necessarily occur every second. Using this scheme introduces the need to buffer data at the transmitting station (aircraft). Eight slots is the minimum number of slots that can be used per burst to a avoid data backlog in the system.

Each 31.5 kHz Iridium channel is divided into eight time slots available for data transmission. These eight time slots are available inside of a 90 ms frame. Therefore, in each second, there are 11.111 frames (1/90 ms = 11.111), meaning that the total number of slots is approximately 88 (11.111 frames per second × 8 slots per frame = 88.88) slots. We will work with the assumption that there are 88 slots available each second. In the one dedicated channel per plane scheme, a plane transmits in its assigned slot(s) every single frame for as long as the plane is using the channel. This fixed slot-burst assignment differs from the dedicated channel assignment technique in that slots used for each transmission are not used on a per-frame basis. That is, the transmission only lasts as long as the slots assigned to it. Due to the variable number of slots that will be used for transmitting, planes can now send different amounts of data. This allows us to reduce (or change) the frequency of each plane's transmissions and introduce a buffer time.

The number of slots used per transmission has an effect on both the number of aircraft that can use a single channel and the required buffer time for each transmit station. The minimum number of slots required for a single transmission is eight slots. In eight slots, 214 × 8 slots = 1712 bits can be transferred, and because each plane generates 1536 bits per second, the eight slots are sufficient to transfer one seconds worth of aircraft data. Also, these eight slots can only carry one seconds worth of data at a time, and therefore each plane would need to use eight slots every second and no buffer time besides the one second required to generate the data (which is not considered delay). As mentioned before, we operate with about 88 slots in each second provided by a single channel. With each aircraft utilizing eight slots per transmission, for example, means that the number of aircraft that can share a single FDMA channel is now approximately 11 (88 slots/8), an increase from eight planes per channel. These eight slots would last 66.24 ms, be able to send 1712 bits, and be accessed in a non-grouped order or in a row. The only requirement is that, in one second, each aircraft would utilize eight time slots. This entails an average transmit rate of 1712 bps per plane. Given the maximum transmitting capability of the system, the number of planes that can be supported using this method can be calculated to be 13,559,040 bits of offered traffic/1712 ≈ 7920 planes. The eight-slot implementation provides an improvement from single dedicated channel use, but is somewhat inefficient because the transmit rate (provided by using eight slots) is larger than the number of bits that we are trying to send. The excess transmission potential is waste. Figure. 4.15 shows the implementation for eight slots per plane.

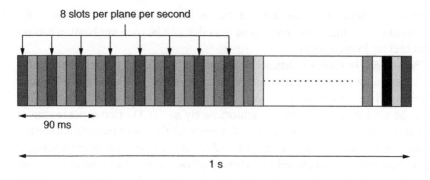

Each aircraft's data is represented by a different color

Figure 4.15 Iridium eight slots per aircraft per second TDMA scheme.

4.5.1.1 Slots per Burst Data Transfer

Buffer Time   The more slots used for each transmission, the more data that can be carried. This increased availability of capacity means each aircraft can use the transmission to send more than one second worth of data. For example, 15 slots per transmission means each transmission can transfer 3210 bits (15 slots × 214 bits per slot = 3210). With each plane generating 1536 bits per second, we can use these 15 slots to send 3072 bits of aircraft data, or two seconds worth of data (1536 × 2 = 3072). The plane is therefore required to be idle for two seconds, buffer data, and use 15 slots in the third second to transmit the data. Buffer time is calculated by dividing the number of bits that can be transmitted in each burst by the number of bits generated by one plane in one second. Figure 4.16 shows the required buffer time for different slots per burst.

Number of Planes per Channel   With each different slot allocation, a buffering time is introduced. When a buffering station is inactive or not transmitting, the channel can be accessed by another plane. Using the example of the 15 slot assignment: with 88 slots per second, and 176 slots in two seconds (88 × 2), 11 (176/15) different and complete 15 slot transmissions can be sent. This division leaves 11 slots that are not used by the 11 planes. These 11 slots can only be used for one second of aircraft data and therefore cannot be used every two seconds. These slots can be configured for re-transmissions, emergency transmissions, or control data. Each of these 15 slot transmissions will contain two seconds worth of data for each plane, meaning that 11 planes can share the channel in two seconds. Figure. 4.17 shows the number of planes that can share a single channel for a specific number of slots per burst.

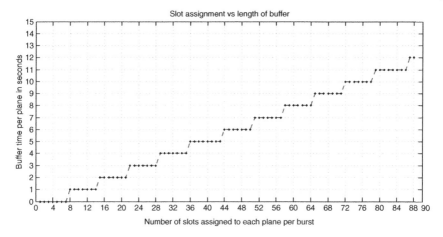

Figure 4.16   Required buffer time for different slots per burst.

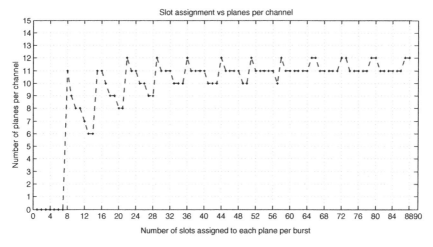

Figure 4.17   Number of planes that can use a single channel.

Transmission Efficiency   The efficiency of each of the different time slot allocation formats depends on how many seconds of aircraft data are sent in each burst and how many bits the allocated time slots can actually carry as illustrated in Figure 4.18. The less efficient allocations use more system capacity than necessary for transferring the aircraft load. The more efficient slot allocations minimize the amount of data carrying capacity of the system not used for data transfer.

Number of Planes That Can Be Supported by System   Each of the different slot assignments offers different capacity management. One of the more important issues to

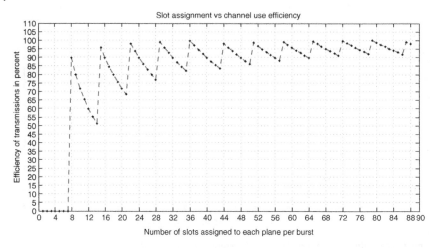

Figure 4.18 Efficiency of each burst (user data/transmission capacity).

consider when implementing a system like this is the minimizing resources used as well as minimizing spectral waste. The spectrum owned by Iridium would need to be purchased, possibly on a per channel or per frequency basis, making it so that using the system that can accommodate the most planes in a fixed band would be most appealing. The number of planes that can be accommodated by the entire allotment of simplex slots available in the system is an essential measure as how well the system is implemented. The entire number of channels is not available to be used for data transfer, but we can see the capacity in terms of aircraft that the channel pool affords. For the most transmission-efficient scenarios in the variable slots per burst scenario, Figure 4.19 shows the number of planes that can utilize the entire pool of channels by the Iridium system.

Slot Use Efficiency In allowing each plane to transmit in a specific number of time slots per specific time period (buffer period), we may see a small percentage of slots that cannot be used at all. These slots will still be available and can be configured for a variety of uses, including retransmission of corrupted data. It is worth noting how many slots will go unused per buffer time as this can also influence our decision on the most efficient technique for data transfer. If we use the 79 slots per burst scenario, we can show the number of slots that are not used for data transfer. The buffer time per plane for the 79 slot scheme is 11 seconds. 12 planes can access the channel at this time and transmit 79 slots each (carrying 11 seconds worth of aircraft data in each burst). 12 planes $\times$ 79 slots per plane = 948 slots that are used in these 11 seconds. With 88 slots occupying 1 second, 11 seconds provides 968 slots. The 948 slots used out of 968 possible in the time frame (20 slots extra) is what we refer to as the slot use efficiency. The extra time slots cannot be carried

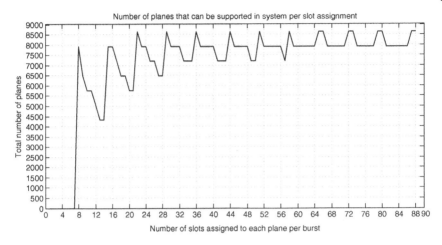

Figure 4.19 Number of planes that can be supported in system.

over to the next bursts because of how time sensitive this particular application is. It would be four more periods of 11 seconds (44 seconds) before enough time slots ($4 \times 20 = 80$ slots) are available to transfer only 11 seconds of data from one plane. For the slot allocation schemes with the highest transmission efficiencies, we derive the Slot Use Efficiency (Table 4.3).

Table 4.3 Comparison of most efficient slot per burst transmissions (Slot efficiency).

		Slots use efficiency			
Slots per burst	**Buffer (s)**	**Slots available**	**Slots used**	**Unused slots**	**Efficiency (%)**
8	1	88	88	0	100
15	2	176	165	11	93.75
22	3	264	264	0	100
29	4	352	348	4	98.86
36	5	440	432	8	98.18
44	6	528	528	0	100
51	7	616	612	4	99.35
58	8	704	696	8	98.86
65	9	792	780	12	98.48
72	10	880	864	16	98.18
79	11	968	948	20	97.93
87	12	1056	1044	12	98.86

Table 4.4 Comparison of most efficient slot per burst transmissions.

	Slots per burst characteristics			
Slots/Burst	Planes/Channel	Buffer(s)	Transmission efficiency (%)	Total planes
8	11	1	89.72	7920
15	11	2	95.70	7920
22	12	3	97.88	8640
29	12	4	99.00	8640
36	12	5	99.69	8640
44	12	6	97.88	8640
51	12	7	98.52	8640
58	12	8	99.00	8640
65	12	9	99.38	8640
72	12	10	99.69	8640
79	12	11	99.94	8640
87	12	12	99.00	8640

Table 4.4 summarizes the most efficient slot assignment bursts. The table includes the number of slots used in each transmission and the number of planes that can use the system with that particular slot per burst arrangement.

4.5.2 Single Second Bursts with Variable Slot Assignment per Frame

A second alternative transmit technique that can be employed is to assign a number of transmit slots for each aircraft for one entire second, once every so many seconds, a time determined by how much data that one second of burst can carry. A single slot per frame is capable of achieving a per second bit rate of 2354 bps, and is hence able to support only one second's worth of aircraft data. However, two Iridium slots per frame, for example, are sufficient to carry three seconds worth of data generated by each aircraft in a one second burst. In two Iridium time slots, 2354 bps × 2 slots = 4708 bits can be transferred. This capacity is enough to support three seconds worth of aircraft data produced at 1536 bits per second 1536 × 3 = 4608 bits. One buffer and burst format that can be used is to require each aircraft to hold a three second buffer, allow it to use two time slots in the following second, and hold its transmissions for another three seconds while it buffers data, and so on. It will be necessary in this scheme for users to use consecutive slots and that the designated number of slots be used in each frame for one second. The quantity of aircraft data that can be sent in each burst of variable slot assignment in one second determines the amount of time that the aircraft buffers the data before using the one second to transmit. Using two slots per frame, for example, means that each aircraft now has an average per second bit rate of 4708 bits per 3 seconds/3 seconds $\approx$ 1570 bps. Figure 4.20 shows the

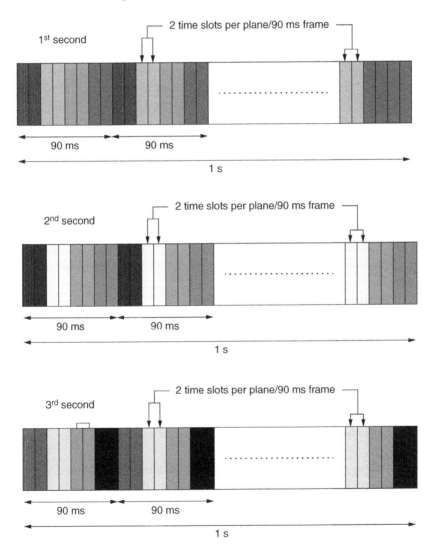

Figure 4.20 Single channel TDMA scheme for three second buffer and burst channel utility.

Table 4.5 Comparison of slot per frame transmissions.

	Slots per frame comparison			
Slots/Frame	Planes/Channel	Buffer (s)	Capacity utilization (%)	Total planes
1 slots	8	1	65.25	5760
2 slots	12	3	97.88	8640
3 slots	10	4	81.56	7200
4 slots	12	6	97.88	8640
5 slots	11	7	89.72	7920
6 slots	12	9	97.88	8640
7 slots	11	10	89.72	7920
8 Slots	12	12	97.88	8640

TDMA scheme for the 3 second buffer and burst channel utility. A one second burst with two slots in each frame using this technique can increase the number of aircraft that can use the system in the mentioned area with four satellite coverage from 5760 (one slot per frame bursts) to about $13,559,040/1570 = 8636$ aircraft. Table 4.5 presents the utilization of capacity across various slot/frame values.

4.5.2.1 Single Second Burst Data Transmission

Buffer Time Buffer time is determined by the amount of data that can be carried in each burst. One slot per frame bursts will support one plane per slot because, the bit rate for one slot per frame is 2354, and only one second of data generated by the aircraft (1536) can be supported. Therefore, each plane transmits every second. With larger slot allocations, more data can be transmitted; therefore, each one second burst can transfer more than one second of aircraft data. This allows each plane to sit idle for a designated number of seconds while it buffers data. The required buffer time for different slots per frame is shown in Figure 4.21.

Number of Planes per Channel Whenever a buffer and burst scheme is used, not all transmitting stations are active at all times. This allows the channels to be shared by multiple users. Figure 4.22 illustrates the number of planes that can use a single channel for different values of slots assigned to each plane in one frame per burst.

Efficiency Is an important aspect to be considered in determining the slot allocation scheme that utilizes the resources to their fullest achievable potential.

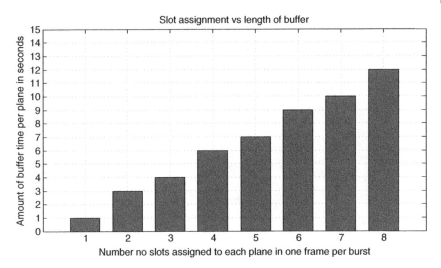

Figure 4.21 Required buffer time for different slots per frame.

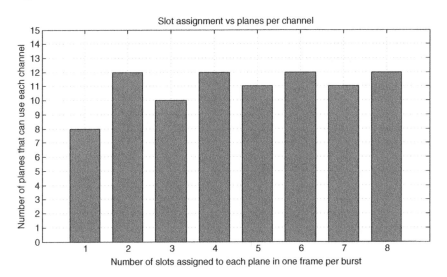

Figure 4.22 Number of planes that can use a single channel.

It is a measure of how much of the available capacity is utilized for user data transmission as illustrated in Figure 4.23.

Number of Planes That Can Be Supported by System When assessing the channel requirement for a fixed number of planes, we are assuming that the area that these aircraft are operating in is covered by four Iridium satellites. These four satellites

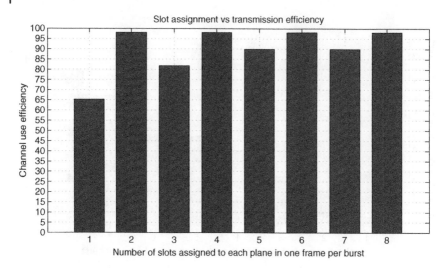

Figure 4.23 Efficiency of each burst (User Data/Transmission Capacity).

supply a total of 720 channels on which communication can be carried out. The total number of planes that can use this amount of resources can be calculated by multiplying the number of planes that use each channel in the different schemes and implementations by 720 available channels (Figure 4.24).

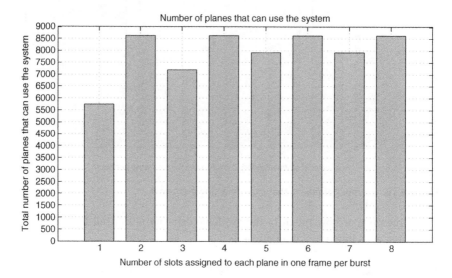

Figure 4.24 Number of planes that can use a single channel.

Slot Use Efficiency In this implementation of channel sharing, channels use their assigned slots for an entire second at a time. Therefore, there are no slots that are unused in this scheme. The slot usage for each of the bursts is 100%.

4.6   Data Transfer – Internet Protocol over Satellite Link Data Transmission

It has been mentioned previously that the Iridium system was designed primarily for two-way voice communications. The majority of the channels available in the system (81.25%) are reserved for voice data while 18.75% of the spectrum is designated as data carrying capacity. In the implementations discussed thus far, the data carrying schemes for Iridium continue to employ user slots for transmission. In this section, we introduce another scheme (Figure 4.25) for the data communication spectrum for the Iridium system. We attempt to use the channel as an open medium available for bursts of varying length unrestricted by time slots. This causes us to employ the use of encapsulating data in packets, in a fashion familiar to Internet protocols. The Iridium spectrum will become a medium for transmitting internet packets from aircraft to the ground and through networks on the ground before being routed to their final destination server.

We concern ourselves mostly with the network and data link layers of the transmission network. That is, we deal mainly with packet formation at a network level and the framing of the packets for transmission at the data link level of the

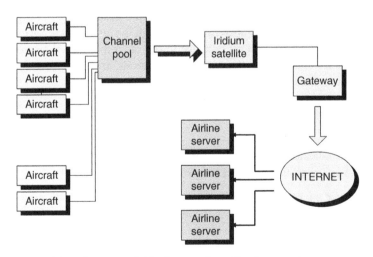

Figure 4.25   System model for Internet-based implementation.

Iridium channel. For simplicity, this data transfer will employ User Datagram Protocol (UDP) at the transport layer.

4.6.1 The Iridium Data Channel

The QPSK modulation used by the system offers a data rate of 50 kbps inside each 31.5 kHz band channel. When considered at a per slot level, each 8.28 ms time slot is able to maintain a data rate calculated as (8.28 ms/90 ms) × 50 Kbs = 4600 bps. When the eight slots per channel TDMA scheme is removed from consideration, the time used by the eight slots allows us to have a channel that can offer a data rate of ((8 × 8.28 ms) 90 ms) × 50 kbps = 36.8 kbps. This raw data rate for the channel is not inclusive of FEC. For fair comparison and consistency, the same 3/4 rate FEC bits will be added to each of the bursts in this data channel implementation.

4.6.2 Packet and Frame Structure

The Transport layer of the system will employ UDP, which is a simple message-based connectionless protocol. UDP has no built-in functions to deal with undelivered or corrupted data reception [Kurose and Ross, 2001]. UDP is selected because of its lightness and connectionless architecture, which allows data transmissions from place to place without network handshakes and connection maintenance. It is also used because of the tight restrictions that will regulate the access to the medium for specific users, eradicating the need for congestion control measures that are included in a more complex and bandwidth consuming TCP message. TCP may however be used after the data has been received at the gateway for more efficient and controlled routing through the internet. Messages sent with UDP are also not guaranteed to arrive in the order in which they are sent. UDP headers are appended to the data at higher levels, and these headers are 64 bits long [Kurose and Ross, 2001].

The UDP header consists of the following fields:

- Source Port Number – 16 bits;
- Destination Port Number – 16 bits;
- Length – 16 bits, indicates the length of the entire datagram;
- Checksum – 16 bits, used for error checking for the header and data.

The network layer packet structure similar to that employed by the Internet version 4 protocol (IPV4) will be used for the analysis. These packets include a header and a body, which will be filled with the user data, in this case, black box flight data. The use of packets for data transfer allows us to address our data at the network layer. The previous schemes of data transfer involved addressing the data at the data link layer with physical addresses. Since the data was transmitted in the same manner as telephone voice data, the address used in the previous schemes was similar to a telephone number on a telephone network. The use of network-level addressing will assist in multiplexing the channels for use by many different planes and/or operators at once based on network addresses. Following the IPV4 packet structure, we propose using packets with the following fields [Kurose and Ross, 2001]:

- *Version*: 4 bits are used to identify the version of the protocol in use.
- *Header length*: 4 bits to show how long the header is.
- *Type of service*: 8 bits long identifying type of transmission.
- *Total length*: 16 bits to indicate the length of the total packet.
- *ID*: 16 bits identifier used to help assemble data at the receiver.
- *Flags*: 3 control bits used to deal with the fragmentation of the packet.
- *Frag offset*: 13 bits to identify the position of the packet in case of fragmentation.
- *Time to live*: 8 bits used to countdown how long a packet is kept in a network.
- *Protocol*: 8 bits identify next-level protocol.
- *Header checksum*: 16 bits used to ensure correct reading of header data.
- *Source add*: 32 bit address to identify the source aircraft of data.
- *Destination add*: 32 bit address to identify the destination of each packet.

Both the transport and network layers add their headers to the data, creating the internet packet (Figure 4.26). This packet is ready to be sent to the data link layer in preparation for transmission to the Earth via the satellite, and eventually routed around the Internet, bound for its destination on the storage server on the ground.

The data link layer for Iridium operates much the same as with the previous schemes. Each burst will have 88 bits of framing and addressing, including sync bits and ramp up/ down power time. As stated before, the system will append a 3/4-rate FEC code to each burst for error checking and correction as illustrated in Figure 4.27.

Once the data is encapsulated by the packet and the frame is retrieved at the Iridium gateway, the data link frame is removed, leaving only the IP protocol packet. The packet is reframed by the data link frame used on the ground level

Figure 4.26 Packet formation for Internet protocol.

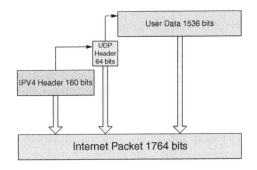

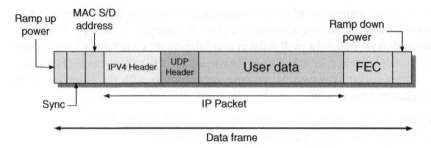

Figure 4.27 Structure of each burst frame.

(such as Ethernet), and the data is then routed through the Internet to the destination address, which is the airline server or ground storage and analysis facility. The implementation of the Internet protocol is not the ultimate goal of the work and is hence not discussed in detail.

The differences in this implementation are simple but important. Firstly, the previous implementations transmitted raw data in the form of bits over an Iridium "phone" data line, while this implementation calls for using internet protocols to deliver data over the lines. The use of time slots in this implementation are replaced by an open data channel, with aircraft transmitting bursts of data. Each of the aircraft in this implementation as well as the home servers for airlines must be assigned some kind of network address similar to an internet address for routing purposes, while the address in the previous schemes was similar to a telephone number.

4.6.3 Data Transfer with Internet Protocols

We now begin looking into how the capacity of the system between the aircraft and the satellites can be achieved. As has been the case, each aircraft generates 1536 bits of data that need to be transmitted to the ground station via the satellite system. We have thus managed to show how the data is prepped for transmission as internet data over our Iridium channel. Our channel has also been altered to become more accommodating to the transfer of data, specifically internet data. This channel offers one slot of 66.24 ms inside a 90 ms frame, which can be accessed by the aircraft in given specified order. This transmission period is available by removing the time slots inside each frame. The 90 ms TDMA frame is still intact and the 23.76 ms in each frame is still used as before for signaling and system functions. Transmission in this time frame allows for data rates up to 36.8 kbps.

4.6.3.1 Setup and Control

The setup and control issues dealt with in this scheme are handled in much the same way as the setup for the schemes already discussed. Aircraft will be required to be connected to a channel before takeoff. The plane will be connected by maintaining and keeping in touch with the available satellites using the control slot available. The satellite allows planes to connect to each of the channels based on how many planes are allowed to use each of the channels and allows no more than the maximum number to connect to each channel. Once the connection is established, the plane is updated with the transmit frequency (how often it will be transmitting) as well as the time that it will be transmitting. The position and travel of each of the planes are monitored by the system in order to be able to handle hand-offs. The plane remains tethered to the channel until either the satellite moves out of view or the plane travels outside of the satellite coverage. Once the aircraft establishes its channel and transmit sequence, it can begin transmitting data, soon before takeoff.

4.6.3.2 Data Packet Transmissions

With each second's worth of aircraft data requiring 2298 bits per transmission (or per second), we derive how many planes can be supported in the system.

Each of our 2298 bit bursts can be transmitted in 36.8K/2.298K = 62.45 ms. If each plane needs to transmit one burst every second, then we can see how 1 second / 62.45 ms = 16.012 transmissions can be sent in one second. This means that 16 different planes can send transmissions in one second on each channel. This is a very significant increase in the number of planes that can transmit on each channel. Removing the restriction of the length of each transmission on each channel creates a freedom that reduces the waste of transmission resources, while allowing the maximum number of bursts in each second, as opposed to the eight bursts per 90 ms frame. Again, we assume that acknowledgments are handled by the system in the simplex and framing time slots provided in each frame.

Using a single IP packet per plane, we see that each of the data channels can be used by up to 16 different aircraft. The theoretical maximum number of planes that can be supported by the four available satellites in our area of interest is calculated by first finding the maximum bit rate offered by the combination of the data channels in the four channels. 720 FDMA channels are designated as data channels by the Iridium system. Each of these channels with the rearranged slot design offers a bit rate of 38.6 kbps. The total bit rate from the combination of channels is therefore 720.6 kbps = 27,792 kbps or 720 × 27.6 kbps = 19,872 kbps when the FEC bits are taken into consideration. In an ideal system that transmits 1536 bits per plane on these channels, the theoretical maximum number of planes we can accommodate is calculated as 19,872 kbps/1.536 kb ≈ 12,938 planes. Obviously, with the use of the IP packet format for data transmission, the number of planes in the system is reduced because of all the overhead data that is transmitted along with each plane's data. With our single packets having a length of 2298 bits including the required FEC bits, we claim that the number of planes that can be supported in this area by the system becomes approximately 12,092 planes (27,792 kbps/2.2298 = 12,093.995). This is a major improvement over any of the other implementations discussed in this paper. The planes using this scheme use the channel once every second, and hence there is no delay. Data is delivered to the satellite from each plane once every second.

The Maximum Transmit Unit (MTU) is the maximum allowable amount of data that the link packet can carry [Kurose and Ross, 2001]. The MTU is a limit on the length of an IP datagram, and any datagrams larger than the MTU will be required to be broken up into fragments or smaller size datagrams before using the link. Fragmented packets require reassembly at the receiver. Using the simple UDP in the transport layer, we would like as much to avoid fragmenting data packets, and as a result, we would like to keep each packet smaller than the MTU. In many networks, the MTU is fixed at about 1500 bytes, or 12,000 bits. With headers adding up to 224 bits (160 bit IP header and 64 bit UDP header), each of our transmissions will be allowed to have a maximum length of 11,776 bits of user data, which is equal to about six full seconds of aircraft data. Therefore, the maximum buffering time for each aircraft in this implementation is seven seconds. Any transmission of packets containing more than six seconds worth of aircraft data will need to be broken into two or more datagrams.

4.7   Number of Channels Needed to Support 5000 Planes

Assuming the average number of planes needed to transfer data to the ground station to be 5000 in every area covered by up to four satellites, we now look at what capacity would be needed to facilitate our different implementations for the 5000 aircraft. The different ways that aircraft share or use channels have been shown, and we now focus on the percentage of the system capacity that will be needed to carry the load for 5000 aircraft at a single time.

In the first scheme that employs assigning a dedicated channel to each transmitting aircraft, 5000 aircraft would clearly need 5000 data channels out of the available 5760 data channels, or 5000 / 8 planes per 31.5 kHz channel = 625 entire 31.5 kHz channels out of the 720 available for data transfer from four satellites. In other words, eight aircraft can utilize each 31.5 kHz channel. This technique ensures that all data is transferred from each of the planes, but it is extremely inefficient. This is due to the fact that each of the channels has the ability to transfer more data than the aircraft are actually sending, rendering the approach quite wasteful of resources. About 35.4% of the system's capacity is wasted this way.

The second scheme involves assigning each aircraft a specific number of slots in which to transmit data in every burst. There are eight time slots in each 90 ms frame in the Iridium TDMA scheme. There are therefore $1/90$ ms ≈ 11 frames per second (or approximately 88 time slots in each second). We have seen that a variable number of aircraft will be able to share each channel of the FDMA channels. This technique shows a significantly improved efficiency from the dedicated channel implementation and does not waste as much of the system capacity. All data is transferred in this technique with some delay as each aircraft buffers data depending on how many slots it uses per burst.

The single second burst scenario is also shown to improve the efficiency of data transfer. With each aircraft using more time slots in each FDMA channel for one second every so many seconds, a different number of aircraft are allowed to share 1 FDMA channel.

The Internet-based implementation of the system is by far the most efficient of all the schemes. One reason for this improvement is the opening up of the FDMA channel to be an open data channel without time slot restrictions. This allows us to use bursts that are not restricted to one time slot. The removal of time slots also allows for more freedom in multiplexing the channel. With the internet-based scheme, 313 FDMA channels are sufficient enough to handle the data of 5000 planes. This is about 44% of the available channels in the area provided by our four satellites.

Table 4.6 shows how the slots per frame and the slots per burst techniques use the least resources of the three schemes. The number of slots per buffer transmission, as well as the slots per frame in one second method both show an improvement over the original scheme of a single dedicated assignment approach. The IP-based system is the most efficient of all the schemes discussed, using the least number of channels. The four satellites that cover our test area of the North American continent provide 720 FDMA channels for data communications, and the 5000 aircraft accomplish data transfer using only about 417 of these using the buffer and burst schemes, which is the second most efficient scheme. The IP-based

Table 4.6 Satellite implementation spectrum use comparison.

Satellite channel requirement (for 5000 planes)			
One channel/Plane	8	625	$625/720 = 86.81\%$
Slots per Burst	12 (max)	416.66	$416.66/720 = 57.87\%$
Slots per Burst	6 (min)	833.33	$454.54/720 = 115.74\%$ (N/A)
Slots per Frame	12 (max)	416.66	$416.66/720 = 57.87\%$
Slots per Frame	8 (min)	625	$625/720 = 86.81\%$
IP-based transmissions	16	313	$313/720 = 43.40\%$

scheme accomplishes data transfer using 313 FDMA channels, while the one channel per plane scheme, the least efficient, would require 625 channels.

4.8 Expected Availability of Spectrum

The Iridium system currently has an estimated 320,000 mobile phone customers. These customers use their mobile devices primarily for voice communication, using the dedicated voice channels offered by the system. The US military is also a major user of the Iridium satellite system, utilizing both data and voice capabilities. As shown earlier, the system channels are divided into FDX voice channels and one way simplex data channels. The entire system has the capacity to offer 64,500 of these simplex data channels globally. These channels are available around the entire globe and are offered by the entire constellation of satellites. There are 15 FDX data channels (30 one-way simplex channels) offered by a single satellite beam, and 48 beams from each satellite, meaning that each of the satellites will be able to provide 720 FDX channels, or 1440 one-way data uplink channels (slots).

Most of Iridium's spectrum is used for voice traffic, and we assume that the one-way data channels are not occupied for long periods of time. Information about the rate of accessing these data channels was not found and assumptions will have to be made concerning the channel's availability. First, we assume that these channels will have to be bought by individual airlines and be guaranteed to be available. Secondly, we will assume that at any given time at least 3/4 of Iridium's data channels are unused by other users. This means about 540 FDMA channels are available for aircraft to use from four satellites. This number is good enough to support 5000 planes using either of our best-case buffer and burst techniques, but not the single dedicated channel method. The 540 channels will be able to support 6480 aircraft with the alternative methods, which is in the ballpark of the peak aircraft traffic over the United States of America.

4.9 Emerging LEO Satellite Constellations

In recent years, the concept of using mega-constellations composed of thousands of LEO satellites to provide global Internet services has become increasingly popular. Commercial enterprises such as SpaceX, Amazon, and OneWeb have all entered the competitive "NewSpace" race to develop and deploy such networks. By integrating LEO satellite networks (SNs) with existing terrestrial networks (TNs), these networks have the potential to provide users with high-bandwidth and low-latency Internet coverage across the globe, even in areas where reliable access has historically been limited [Zhang et al., 2022]. However, early LEO constellations such as Iridium and Globalstar were relatively small, with only around 100 satellites. This limited the number of satellites within the line-of-sight of each ground station, which constrained the satellite-ground topology design space. Additionally, the impact of the satellite-ground topology was limited to just one hop between the satellite and the ground station, which resulted in a limited set of design criteria [Lai et al., 2022]. As LEO constellations have grown in size, with thousands of satellites now in orbit, the number of visible satellites in ground stations has greatly increased, particularly in low and middle latitudes. This has significantly extended the design space for satellite-ground topology, requiring ground stations to choose which satellite to connect with. The larger constellation scale presents exciting new opportunities to transform the global Internet landscape, but it also requires the development of new technologies and algorithms to optimize network performance. In this context, an algorithm is proposed and implemented to enable aircraft to select the best line-of-sight satellite to minimize overall system latency.

4.9.1 Problem Formulation

The problem at hand is to develop a real-time communication system that can connect airplanes, satellites, and ground stations in an efficient manner. The primary goal of this system is to reduce the average latency or hop count of the communication network. However, this objective needs to be achieved while ensuring that several constraints are met. The first constraint is that links between airplanes, satellites, and ground stations can only be established when they are visible to each other. This means that a visibility matrix must be used to determine when communication links can be established. The matrix takes into account the location and movement of each element in the system to determine when they are within line-of-sight. The second constraint is that the average topology change interval between all airplane-ground station pairs must be greater than a certain threshold. This is important for ensuring that the communication network is stable and reliable. The system must be able to maintain connections between airplanes and ground stations even as they move around. The third constraint is that each airplane can only establish one airplane–satellite link, while the maximum number of inter-satellite links is four (as in the case of Starlink). This ensures that the communication network is not overloaded and that communication resources are used efficiently. The fourth constraint is that the Dijkstra algorithm's minimum hop count must be used to determine the optimal path between any airplane-ground station pairs. This ensures that the communication network is optimized for minimal latency. The fifth and final constraint is that the minimum airplane–satellite link data rate should be at least 1800 bps. This ensures that the communication network can support high-bandwidth applications and services. Below are the summarized constraints that need to be taken into consideration in designing a real-time communication system for airplane-satellite-ground station interconnection to achieve the objective of minimizing average latency:

1. Links can only be established between airplanes, satellites, and ground stations when they are visible. This requires the use of a visibility matrix that takes into account the location and movement of each element in the system.
2. The average topology change interval between all airplane-ground station pairs must be greater than a certain threshold. This is necessary to ensure a stable and reliable communication network.
3. Each airplane can establish only one airplane-satellite link, while the maximum number of inter-satellite links is four (as in the case of Starlink).
4. The Dijkstra algorithm's minimum hop count must be used to determine the optimal path between any airplane-ground station pairs.
5. The minimum airplane–satellite link data rate should be at least 1800 bps.

To address these constraints, a proposed algorithm has been developed. The algorithm has several steps, starting with connecting to VDL4 at takeoff and landing. VDL4 is a communication standard used in the aviation industry for ground-to-air and air-to-ground communication. After takeoff, the algorithm calculates the minimum remaining service time requirement of satellites to ensure that the airplane is connected to a satellite that has sufficient service time to meet its communication needs. The algorithm then selects a candidate satellite set for the airplane that meets the remaining service time requirement and moves in the same direction (clustering). This step ensures that the airplane is connected to a satellite that is moving in the same direction, which reduces latency and improves communication quality. An inter-satellite algorithm is applied to the ground station, taking into account the topology of the satellite network, to ensure that the airplane is connected to the ground station through the optimal path with the minimum hop count. The proposed algorithm ensures that the communication system is stable, reliable, and provides low-latency communication between airplanes, satellites, and ground stations. To validate the proposed algorithm, simulations were conducted using the Matlab Aerospace Toolbox to simulate the flight trajectory of an aircraft from Philadelphia to New York (a 2.5-hour flight) along with the Starlink constellation. The proposed algorithm's latency analyses are currently being worked on to further validate its efficacy.

Proposed Algorithm for Airplane-Satellite-Ground Station (GS) Interconnection:

1. At takeoff and landing, connect using VDL4, which is a communication standard used in the aviation industry for ground-to-air and air-to-ground communication.

2. After completing takeoff, calculate the minimum remaining service time requirement of satellites. This is necessary to ensure that the airplane is connected to a satellite that has sufficient service time to meet its communication needs.
3. Select the candidate satellite set for the airplane that meets the remaining service time requirement and moves in the same direction (clustering). This step ensures that the airplane is connected to a satellite that is moving in the same direction, which reduces latency and improves communication quality.
4. Apply an inter-satellite algorithm to the ground station. This algorithm takes into account the topology of the satellite network and ensures that the airplane is connected to the ground station through the optimal path with the minimum hop count.

    The proposed algorithm ensures that the communication system is stable, and reliable, and provides low-latency communication between airplanes, satellites, and ground stations.

4.9.2 Results

In order to validate the algorithm mentioned earlier, we employed the Matlab Aerospace Toolbox to simulate the flight trajectory of an aircraft from Philadelphia to New York (a 2.5-hour flight) along with the Starlink constellation. Next, we visualized the path between the aircraft and a space-based ground station through the satellite constellation.

Figure 4.28 provides a visual representation of the proposed scenario and the flight trajectory of the aircraft. However, it is important to note that the proposed algorithm has yet to undergo a thorough analysis of its latency performance.

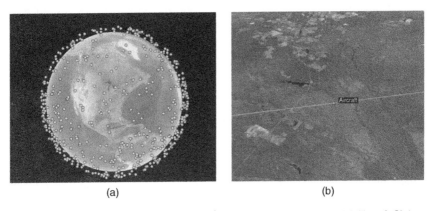

(a)   (b)

Figure 4.28 (a) Snapshot of the proposed algorithm implementation. (b) Aircraft flight trajectory snapshot.

This means that while the algorithm shows promise in minimizing the average latency and satisfying the various constraints, its actual latency performance remains to be evaluated. Further analysis and testing will be required to determine the effectiveness of the proposed algorithm in achieving its latency-related objectives.

4.10 Discussion

It can be seen from the work in the chapter that the different channel accessing or sharing schemes provide some sort of improvement over dedicated channel implementation. The dedicated channel scheme would be the easiest system to implement, as it mirrors more closely, the general use of the Iridium system. The introduction of channel sharing adds complexity to the system, as the number of planes using each of the channels has to be tethered to the channel even when not transmitting. The IP-based technology will require additional hardware and software (modems) and is the most complex to implement for the entire group.

Table 4.3 shows the number of slots used for actual data transmission and how many (if any) are unused. From Table 4.4, we see that there are ten slots per burst slot count can transfer for a maximum of 8640 aircraft at any given instance. These ten slot counts differ in transmission efficiency, slot use efficiency, and buffer time. Buffer time is an important characteristic to be used in determining the most suitable slot assignment per burst, but other issues can be used in making this decision. Some of the assignments use 100% of the available slots for data transmissions, while other scenarios leave some unused slots after data is transferred. These excess slots can be utilized for satellite-to-aircraft communication, by way of retransmission slots for corrupted or lost transmissions. Since the Iridium system was designed primarily for voice communication, provisions for sending and receiving ACKs are not included. With voice data, acknowledgment and retransmissions are ineffective due to timing issues with voice communication. In data transfer, retransmissions and acknowledgments are easier to implement. With this in mind, there are three of the most efficient slot-per-burst implementations that would work best.

The slots per burst designs that would work best are listed here.

- The 15 slots per burst implementation allows 11 planes to use a single channel with 95% transmission efficiency (from slots used in actual data transmission). Each plane stores data for two seconds before using 15 slots to transmit the buffered data. This scheme leaves 11 unused slots every two seconds. With 11 planes connected to the channel, the 11 slots can be assigned to each plane for receiving an ACK from the satellite.

- Transmitting with 65 slots per burst allows 12 planes to utilize a single channel with a nine-second buffer time. This assignment also leaves 12 unused slots in the given time that can be used for the ACK transmission.
- The 87 slots per burst require 12 seconds of buffer time per plane and can provide 12 slots for ACKs.

As mentioned earlier, the work in this book assumes that all data slots available from the system are used for data uplink from aircraft to satellite and that, through inter-satellite routing and networking, the data is relayed to the Earth gateways on other channels. The three slots per burst implementations listed above provide availability for 7920 planes (15 slots per burst) and 8640 planes (65 and 87 slots per burst). The manner in which the satellites relay the data to the gateways is not included in the research.

The internet-based approach utilizes satellite channels in a manner not similar to the other schemes. The use of Time slots is done away with in the channels, and access to the channel is based on the length of transmissions or bursts. Implementing the internet packet-based system for this purpose is easier to design since the number of users of the system are known, or expected as well as the distribution of their transmissions. As a result, issues such as packet collisions and queues can be avoided at this transmission stage. The use of internet protocols requires some additional hardware and software to implement. The system's complexity in total increases with the satellite internet scheme. The internet scheme, with the open access data channel, can be seen to be the most theoretically efficient scheme for our data transfer, allowing the most number of aircraft to use the available spectrum.

We can see that when the planes are allowed to share the channels, the number of channels required to handle the load is reduced. This reduction in number of channels needed comes at the expense of a small delay time at the receiver. Too large delay time is obviously not tolerable for this application, but implementation protocols will have to include an emergency transmit permission for emergency situations, giving priority to aircraft in distress, where data can be streamed using all necessary resources. Satellite communications are not currently used for voice transmissions from aircraft or for any other data transfer from aircrafts, therefore, the available channels used are strictly for black box data transfer.

4.11 Summary

This chapter discussed the potential implementation of ground-based FDR/CVR recorders using the emerging LEO satellite constellations. The proposed algorithm for airplane-satellite-ground station interconnection is introduced with the objective of minimizing average latency while satisfying various constraints such as limited link establishment and the use of Dijkstra's algorithm. However, it is important to note that the proposed algorithm has not yet undergone a thorough analysis of its latency performance. Although the algorithm shows promise, further analysis and testing are required to determine its effectiveness in achieving its latency-related objectives.

References

Aireon. All Things Automatic Dependent Surveillance-Broadcast (ADS-B), 2018. URL https://aireon.com/1090-global/back-basics-part-2-things-automatic-dependent-surveillance-broadcast-ads-b/.

eoPortal. Iridium NEXT, January 2013. URL https://www.eoportal.org/satellite-missions/iridium-next#iridium-next-hosting-payloads-on-a-communications-constellation.

FCC. Iiridium NEXT engineering statement, 2013.

C.E. Fossa, R.A. Raines, G.H. Gunsch, and M.A. Temple. An overview of the IRIDIUM (R) low Earth orbit (LEO) satellite system. In *Proceedings of the IEEE 1998 National Aerospace and Electronics Conference. NAECON 1998. Celebrating 50 Years (Cat. No.98CH36185)*, Dayton, OH, USA, pages 152–159, August 1998. doi: 10.1109/NAECON.1998.710110.

W. Graham. Iridium NEXT-5 satellites ride to orbit on SpaceX Falcon 9, March 2018. URL https://www.nasaspaceflight.com/2018/03/iridium-next-5-satellites-spacex-falcon-9/.

Yvette Hubbel and Lockheed Sanders. A comparison of the IRIDIUM and AMPS system. *IEEE Network Magazine*, 11(1):52–59, 1997.

iridium. History of iridium, n.d. URL https://www.iridiummuseum.com/timeline/.

Mohammad Abdul Jabbar. Multi-link iridium satellite data communications sytem. Master's thesis, Osmania University, Hyderabad, India, 2001.

James Kurose and Keith Ross. *Computer Networking, A Top Down Aproach Featuring the Internet*. Addison Wesley Longman Inc., 2001.

Zeqi Lai, Weisen Liu, Qian Wu, Hewu Li, Jingxi Xu, and Jianping Wu. SpaceRTC: Unleashing the low-latency potential of mega-constellations for real-time communications. In *IEEE INFOCOM 2022 - IEEE Conference on Computer Communications*, pages 1339–1348, 2022. doi: 10.1109/INFOCOM48880.2022.9796887.

P. Lemme, S. Glenister, and A. Miller. Iridium aeronautical satellite communications. *IEEE AES Systems Magaazine*, 14(11):11–16, 1999.

S. Matale. Ground-based black box system implementation using satellite and VHF data link networks. Master's thesis, The University of Mississippi, Oxford, USA, 2010.

Bill McIntosh. Down to earth reasons for iridium failure, August 1999. URL https://www.independent.co.uk/news/business/down-to-earth-reasons-for-iridium-failure-1113638.html.

D. Millard. Iridium: Story of a communications solution no one listened to, August 2016. URL https://www.newscientist.com/article/mg23130850-700-iridium-story-of-a-communications-solution-no-one-listened-to/.

R. Nelson. Iridium: From concept to reatilty. *Via Satellite Magazine*, September 1998. ISSN 1525-3511. doi: 10.1109/WCNC.2006.1696547.

S. Pratt, R. Raines, C. Fossa, and M. Temple. An operational and performance overview of the iridium low earth orbit satellite system. *IEEE Communications Surveys*, 2(2):2–10, 1999.

Spacecraft & Satellites website. Iridium-Next, 2018. URL https://spaceflight101.com/spacecraft/iridium-next/.

Yaoying Zhang, Qian Wu, Zeqi Lai, and Hewu Li. Enabling low-latency-capable satellite-ground topology for emerging LEO satellite networks. In *IEEE INFOCOM 2022 - IEEE Conference on Computer Communications*, pages 1329–1338, 2022. doi: 10.1109/INFOCOM48880.2022.9796886.

5

VHF Digital Link Implementation

This chapter explores communications via a very high-frequency digital link (VDL) and assesses the ability of such a link (mainly VDL mode 4) to transmit crucial flight data. It is important first that we give a solid understanding of the current VDL systems to determine the best performance for transmitting flight data then we will demonstrate that data transfer using the VDL mode 4 system on aircraft is feasible.

5.1 VHF Communications System

Very high frequency, also referred to as VHF, is the frequency range of radio waves from 30 to 300 MHz. This range is used for several aircraft communications. The VOR (VHF omnidirectional radio range) and the instrument landing system (ILS) localizer are deployed in the aviation sector as air navigation beacons on frequencies 108–118 MHz. Aircraft can use VOR to find their way to the ground station. The VOR receiver on board an aircraft can demodulate a signal from a VOR transmitter station to calculate the aircraft's bearing with respect to the station. The plane's location can be determined by triangulating three or more stations. ILS are installed in planes to aid pilots in their approach to the runway and landing. This is particularly important in cases of low visibility brought on by inclement weather, night landings, and crosswind approaches. The ILS localizer helps the pilot to steer appropriately to the left or right during the landing [Bin Rahim and Breuer, 2011]. A localizer transmitter can be found close to the runway's ending. Horizontally oriented antennas transmit two intersecting beams, the received signal will be equally modulated down the runway's centerline. The difference in depth modulation (DDM) allows the detection of any departure from the runway centerline [Bin Rahim and Breuer, 2011].

The spectrum of 118–137 MHz is used as airband (the VHF radio spectrum used in civil aviation) for air traffic control (ATC), and Amplitude Modulation (AM)

Real-Time Ground-Based Flight Data and Cockpit Voice Recorder: Implementation Scenarios and Feasibility Analysis, First Edition. Mustafa M. Matalgah and Mohammed Ali Alqodah.
© 2024 The Institute of Electrical and Electronics Engineers, Inc. Published 2024 by John Wiley & Sons, Inc.

voice communication. As of 2012, most countries divide this band into 760 channels in steps of 25 kHz. In Europe, it is becoming common to further divide channels into three (8.33 kHz channel spacing), potentially permitting 2280 channels [Transport Canada, 2021]. The frequency of 121.5 MHz, often known as the international air distress frequency or guard frequency, is the frequency that is considered to be the standard for civilian aircraft in the event of an emergency. If ATC is unable to make contact with a plane on its primary frequency, ATC will move to the guard frequency in an attempt to communicate with the aircraft. VHF sustain the majority of near-ground–air communications. Due to the use of "Line Of Sigh" transmission, the range is approximately 30 miles for aircraft flying at 1000 feet above ground level and 135 miles for aircraft flying at 10,000 feet [SAMYOG KC, 2020].

The VDL communications system is one of several aircraft-to-ground subnetworks that can be used to provide data communications between aircraft-based application processes and their ground-based peer processes through the aeronautical telecommunication network (ATN). Plain Old ACARS (POA), also known as VDL mode 0, mode 2, which is also used for ACARS, mode 3, and mode 4 are the four distinct versions of VDL that have been developed. In Section 5.2, we will discuss all modes and provide a comprehensive technical description of mode 4 since this is the mode we will use to send data from FDR/CVR recorders.

5.2 VDL Modes

VDL modes are distinguished primarily by their multiple access scheme and modulation technique, with the exception of VDL mode 0, which was identical to VDL mode 2 but it used the same VHF link as VHF ACARS and thus could be implemented using analog radios before the completion of the implementation of VHF digital radio. Table 5.1 outlines the fundamental differences between VDL modes 0, 2, 3, and 4.

Table 5.1 Link characteristics.

VDL mode	Multiple access	Data rate (kbps)	Modulation	Network control	Traffic
Mode 0		2.4	AMSK	Air to ground	Data
Mode 2	CSMA	31.5	D8PSK	Air to ground	Data
Mode 3	TDMA	31.5	D8PSK	Air to ground	Voice/data
Mode 4	STDMA	19.2	GFSK	Air to ground, Air to air	Data

5.2.1 VDL Mode 0

In July 1978, the Aeronautical Radio Incorporated division developed the first data communication system known as ACARS. This new communication strategy lowered staff workload while improving data integrity. This technology, nowadays known as Plain Old ACARS (POA) and still in use in some aircraft, makes use of VHF channels that were originally allocated for voice transmission [Mahmoud et al., 2014]. It uses Amplitude Modulated Minimum Shift Keying (AMSK) to achieve a data rate of 2.4 kbps and was originally used for airline communications. Lately, ACARS has been enhanced, and its use has extended to incorporate multiple means of communication, such as SATCOM and HF links. ATC authorities began to promote the use of ACARS between controllers and pilots in the 1980s to improve air traffic management safety and efficiency [Mahmoud et al., 2014]. This mode, also known as VDL mode A, does not have enough capacity to transmit large amounts of data efficiently, so it will not be considered for flight data transmission.

5.2.2 VDL Mode 2

VDL mode 2 is a data link mode that uses 25 kHz channels in the airband. This mode of operation is designed to support data communications only (no voice) and is a connection-oriented link. Using Differential 8 Phase Shift Keying modulation; each channel can have a theoretical bit rate of 31.5 kbps. VDL mode 2 requires ground infrastructure for communicating between ground stations and aircraft. Additionally, the modulation scheme used in mode 2 has poor performance in the presence of noise and has a high signal-to-noise ratio (SNR) threshold [Matale, 2010]. This sensitivity to noise requires two guard channels per data channel, making this mode spectrally inefficient. Furthermore, the link does not support priority transmission and uses media access based on the ALOHA algorithm, resulting in poor performance when packet traffic increases. This problem makes it unsuitable for transmitting and receiving time-critical data, such as the data generated by the flight data acquisition unit[Bretmersky et al., 2002]. Following is a description of how the VDL system connects to the Open Systems Interconnection (OSI) model's first three service delivery layers.

The VDL system is related to the three lower layers of the OSI model [Bae et al., 2009]. The first layer is the physical layer responsible for modulation, encoding, and forward error correction using interleaving and Reed Solomon coding. The second layer called the Link layer, consists of two sublayers and a link management entity: the Medium Access Control (MAC) sublayer, which handles channel access and offers access to the physical layer using carrier sense multiple access (CSMA) algorithms, and the data link service sublayer, which handles frame exchanges, frame processing, and error detection. Connectivity and maintenance of a data link service (DLS) network's sublayers are both controlled by the link management entity (LME). Finally, the third layer is the network layer, and to be more precise, only the SNAcP network sublayer is considered part of the VDL system. Throughout a virtual circuit, data packets can be delivered and received. Additional capabilities such as error recovery, connection flow control, fragmentation, and subnetwork connection management can be utilized. Message collisions are unavoidable in the CSMA technique used in mode 2, and message delay grows exponentially with the number of available connections. As a result, VDL mode 2 exhibits nondeterministic behavior and cannot be used in time-critical settings [Li et al., 2006].

5.2.3 VDL Mode 3

Using the same D8PSK modulation algorithm as VDL mode 2, data can be transmitted at a rate of 31.5 kbps using VDL mode 3. The TDMA protocol is used to control access to the medium. In addition to data transmission, VDL mode 3 supports simplex voice conversations across the link by providing multiple modes of operation that feature dedicated time slots for voice [Bretmersky et al., 2002]. Since ground stations are required by TDMA technique used in VDL mode 3, it employs the same D8PSK modulation mode 2 uses, giving it the same disadvantage and problems with noise and spectrum waste as mode 2 [Matale, 2010].

5.2.4 VDL Mode 4

To facilitate air–air and ground–air communications and related applications, VDL mode 4 uses a self-organized time division multiple access (STDMA) scheme, setting it distinct from the other VDL modes. Stations (mobile and ground) in the STDMA scheme use a reservation protocol to announce their intent to transmit within a particular time slot before actually sending the message [Li et al., 2006]. To make the most of available data links and ensure that no two stations are sending out messages simultaneously, it is necessary for all adjacent stations, which are separated by only one hop, to be informed of the precise timing and planned use of slots. Another advantage of mode 4 is that data signals to be transmitted in this mode are modulated using Gaussian Frequency Shift Keying (GFSK), which exhibits a superior power efficiency to D8PSK modulation [Matale, 2010]. This means that the required SNR is lower in mode 4, than in modes 2 and 3. Having a lower SNR increases the frequency reuse factor (as the signal to interference ratio is reduced) in mode 4, thereby increasing the capacity of the system. Since this mode is related to the book's purpose, we will elaborate more on its technical specification.

VDL mode 4 is a data communication link used by aircraft to communicate directly with other aircraft or straight to a ground station. The link uses STDMA is a channel access method to accommodate multiple user channels. Gaussian filtered Frequency Shift Keying (GFSK) allows each of the 25 kHz (frequency division multiple access [FDMA]) channels to transmit at a bit rate of 19,200 bps. The available band of frequencies assigned for aircraft communications (118 – 136.975 MHz) is 18.975 MHz wide and can accommodate a total of 760 of these 25 kHz channels. In the MAC Layer, each superframe spans 60 seconds. Design flexibility allows a variable number of time slots to be available inside of the 60 second superframe, with 60 time slots being a minimum number and 15,360 time slots being the maximum. The default setting for number of slots available in a superframe is 4500 [Matale, 2010], and this is the number that will be used in our analysis for this research. With each one minute superframe consisting of 4500 slots, each of these time slot will be 13.33 ms long (60 s/4500 slots per min = 13.33 ms). Transmitting stations request, and are assigned reserved access to time slots in each frame in advance, for periodic or unicast transmissions. The slots can also be designated as fixed access times lots for specific users or stations (such as a ground station). Time slots can also be accessed at random when they are not reserved, or when access is not fixed. Each node accessing a single FDMA channel is allowed to use a maximum of 75 time slots (or one second) per superframe for a single transmission [Bretmersky et al., 2002]. With a data rate of 19,200 bits per second, the raw data rate for each time slot is 256 bits (256 bits per 13.33 ms). VDL mode 4 has the capability of separating data and control onto different channels, meaning that control bits are not necessary in every single time slot of a transmission [Bretmersky et al., 2002]. For example, if each station plans to transmit using four reserved time slots, then control data are only necessary in the first slot. Therefore, the number of information bits that can be sent (on a per time slot average) depends on the overall message length. Regardless of the content of the data (information or control), each time slot will be assumed to have 16 bits of ramp up power stabilization time and 24 bits of ramp down power decay time, as well as a 24 bit synchronization field. Therefore, in a situation where only one time slot is used for transmission, a total (or maximum) of 192 bits (256 − 64) can be delivered (Figure 5.1).

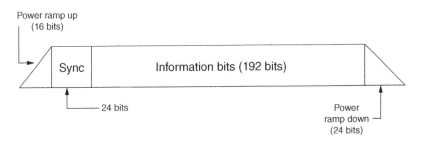

Figure 5.1 VDL mode 4 time slot.

5.3 Data Transfer – VDL Mode 4 Implementation

To facilitate the transfer of all the data generated by the FDR recorder (1536 bps), each plane must use multiple slots inside each superframe. Each aircraft in the system must reserve a slot for future transmitting instead of having a fixed slot assignment because of the dynamic nature of the process. As the aircraft travel, they are assigned new cells to operate, and this cross-over may occur frequently.

Every time slot used for data transmission is required to be reserved in VDL mode 4. A transmitting station begins by sending three requests in three unused time slots, with information used for requesting future data transmitting time slots. The three slots are necessary because each Request To Send (RTS) message is random access using slotted Aloha. These three slot requests are assumed to return one reservation for time slots whenever they are sent. In other words, we work with the knowledge that there is never a situation that results in failure to secure slots for future transmissions. An unused time slot for the reception of a Clear To Send message is required with every three RTS before information transmission can commence. Reservation of an additional slot is required for receiving an acknowledgment (ACK) after the transmission burst is complete. Due to the constant relocation of aircraft, and the foreseeable need to change channels regularly, the reservation for transmission slots cannot be assumed to be a periodic transmit reservation. Therefore additional five-time slots are required for every burst of user information used for our flight data transfer schemes. Figure 5.2 shows the proposed system model for the VDL mode 4 implementations.

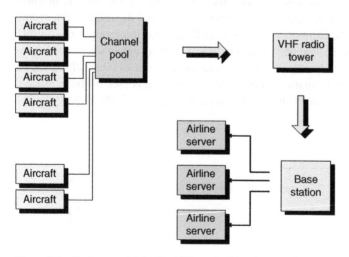

Figure 5.2 System model for the VDL mode 4 implementation.

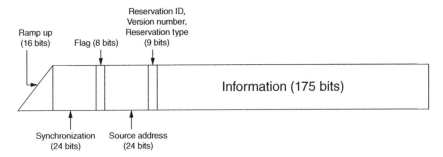

Figure 5.3 First data carrying time slot.

It is important that each plane is capable of transmitting at least 1536 bps. Inside all data bursts, an aggregate number of bits per slot of cyclic redundancy check (CRC) and FEC bits are used to deal with error detecting and correcting at the receiver, and these bits are added in the last time slot of the burst. The first slot of every multiple-slot transmission will need to have 16 bits for power ramp-up, 24 bits for sync, 8 flag bits, 24 bits for the source address, and 9 bits of reservation ID and data. This leaves the first slot with the ability to carry 175 bits of information. Figure 5.3 shows the first time slot.

The second slot (assuming more than two slots per burst), and all other in-between slots, will be able to function with 100% utility, carrying 256 bits of information. The last time slot will need to carry the bits for the CRC bits and 24 bits for the power ramp down. Therefore the last slot in every burst will be able to hold a maximum of 232 bits (for user data smaller than 512 bits), comprising information and error correction and detection. For all larger quantities of data, 64 bits of CRC will be appended to the last slot in the transmission.

Using these estimates, we see that the number of time slots required to hold 1536 bits of data is 7. The first slot will carry 175 bits, and the next five will carry 256 bits. We can assume that for transmission of this size, about 64 error checking and correction bits will suffice, and the last slot will therefore be able to hold 168 bits of user data. Our seven slots can then carry $(175 + (256 \times 5) + 168 = 1623)$. This number is sufficient to provide each aircraft with the necessary capacity for its data load.

5.3.1 Consecutive Time Slot Bursts

Every burst will consist of 3 RTS slots, 1 CTS slot, 7 data slots, and 1 ACK slot, totaling 12 slots. With each time slot occupying 13.33 ms, 12 slots will last $(13.33 \text{ ms per slot} \times 12 \text{ slots} = 159.96 \text{ ms})$. Each aircraft must maintain a minimum of one burst per second to ensure that data is entirely transferred. With each burst lasting a minimum of 159.96 ms, each channel can supply a burst for

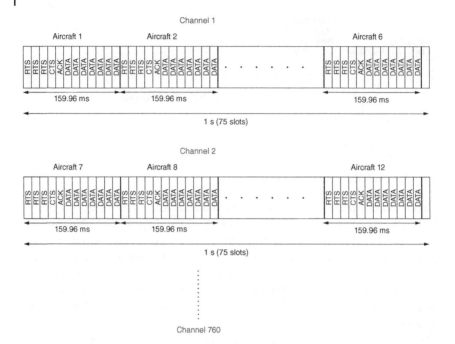

Figure 5.4 Slot usage for data transfer in VDL mode 4.

6 (actually 6.252) planes in each second. Six planes, each utilizing 159.96 ms per second, as could be seen from Figure 5.4, leaves about 40.24 ms or three time slots extra in each second that is unutilized for aircraft data. These slots can be configured for use by base stations to convey data such as handoff information to aircraft using the channel. With this implementation, it is expected that data will be received at the rate of one second worth of data per second.

The spectrum used for VDL mode 4 communications (118 − 136.975 MHz) can be divided into 760 channels of 25 kHz bandwidth. It is each one of these channels that can support six aircraft each, meaning that up to 760 × 6 = 4560 aircraft can be supported at a single time per single cell. It is essential to remember that VDL mode 4 channels can be used over and over again, granted that they are reused at a distance far away enough to not cause interference. With such a large number of channels available over a sizeable area, the available capacity in the system is more than enough to facilitate the data transfer. However, the entire spectrum is not used exclusively for VDL mode 4 and flight data transfer. The other aviation communications systems utilize the same band of frequencies, and there can not be 760 channels available in each cell used only for flight data transmissions.

Assuming that base station receivers will be available at locations such as airports all across land masses, the capacity that is offered by VDL mode 4 is sufficient to carry data efficiently for far more than the average number of 5000 airborne aircraft in the USA.

5.3.2 Alternative VDL Mode 4 Transmission Scenarios

When the entire spectrum for use in VDL mode 4 is assumed to be available, it can be seen that no problems would be encountered with getting aircraft flight data to the ground. However, the availability of these channels is not guaranteed, and we would like to be able to derive maximum capacity estimates for a finite number of channels. We would like to look at situations where we can use the least available channels to carry the data.

In STDMA channel access schemes that do not use periodic reservation transmissions, a large number of time slots are needed to reserve transmission slots. It is this need to use the channels for reservations every burst that makes it counterproductive to transmit a few time slots of data at a time(given that each transmission requires RTS slots). It is wasteful to transmit one slot of data at a time when five additional slots are necessary to set up that transmission. The larger the number of slots used for data per burst, the more efficient the scheme becomes.

5.3.2.1 No Buffer and Burst

This scenario, as we already discussed before, requires each aircraft to transmit one second worth of data every second. In this case, data will be received at the receiver delayed by a maximum of one second (the one second it takes the Flight Data Acquisition Unit (FDAU) to generate the data). One second of aircraft data is 1536 bits long. A load of data requires a minimum of seven data-carrying transmission slots that can transmit 1623 bits (5 slots carrying 256 bits, the first slot carrying 175 bits, last slot carrying 168 bits). The total length of the burst including RTS and Clear to Send (CTS) slots is therefore 12, or 12 slots per plane per second. This burst lasts 159.96 ms. At this rate, one channel can support 6.25 planes ($\approx$ 6 planes), as seen in Figure 5.5.

When buffering, a plane collects data for a predetermined amount of time before being allowed to send the data accumulated in that time period. In VDL mode 4

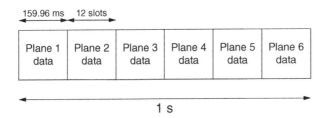

Figure 5.5 Single channel utilization using no buffer.

buffering and bursting are additionally advantageous because each burst of data requires five nondata slots to be used as well as data carrying slots, and so larger chunks of user data utilize the channel more efficiently. In the scheme where each plane transmits once every second, there is no delay in the transmission. We will consider schemes that require each aircraft to store (buffer) data for a period of time, and transmit a larger load at a later time, in an effort to use channels more efficiently. In the following, we will analyze different transmission scenarios to determine the most efficient burst formats that can be used in VDL mode 4 Data transfer.

5.3.2.2 Two Second Buffer and Burst

This scenario requires each aircraft to buffer two seconds worth of data before transmitting data, meaning the received data will be two seconds delayed, and up-to-date data will be received every two seconds. Two seconds of aircraft data are 3072 bits long. Total of 3072 bits of data require a minimum of 13 data carrying transmission slots (11 slots carrying 256 bits, the first slot carrying 175 bits and the last slot carrying 168 bits). The transmission including RTS and CTS would therefore need 18 time slots, that is 18 time slots per plane every two seconds. The total time for each transmission will be (13.33 ms per time slot × 18 time slots = 240 ms). Each plane uses 240 ms every two seconds or 18 timeslots every two seconds, as seen in Figure 5.6. At this rate, each channel can now support (2 s/240 ms = 8.33 planes ≈ 8 planes).

5.3.2.3 Three Second Buffer and Burst

In this scenario, each aircraft buffers up to three seconds worth of data acquired from the FDAU before sending. A minimum of 19 time slots are needed to transfer the 4608 bits stored up after three seconds of buffering. The total number of (19 data carrying slots + 5 RTS CTS ACK slots) 24 slots per burst utilize 320 ms per channel. Each channel can therefore handle 3 s/320 ms ≈ 9 planes per channel, as seen in Figure 5.7.

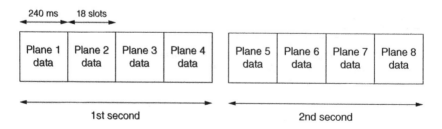

Figure 5.6 Single channel utilization using two second buffer.

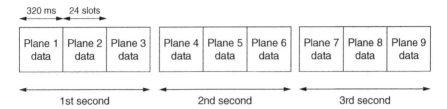

Figure 5.7   Single channel utilization using three seconds buffer.

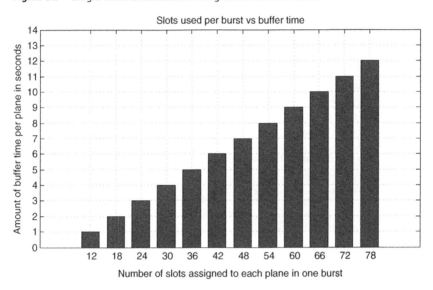

Figure 5.8   Amount of buffer time for different number of slots per transmission.

Following these steps, we can now extend this technique to determine the relationship between the amount of buffer time that each burst of varying slots length, as seen in Figure 5.8.

Figure 5.9 shows the number of planes that can occupy each channel, given the buffer time and the number of slots, used per transmission burst.

The efficiency of these transmissions, shown in Figure 5.10, is a measure of how much of the available resources are used for actual data. The VDL mode 4 transmission is unique in that each burst requires an additional five time slots per transmission, regardless of the number of data-carrying slots used. This makes larger data transfers immediately more efficient than smaller slot utilizing bursts. The graphs below show the percentage of data carrying slots that are used for actual user data, given the number of data-carrying slots, and the percentage of the entire burst (data slots and RTS, CTS, and ACK slots) used for actual data transmission.

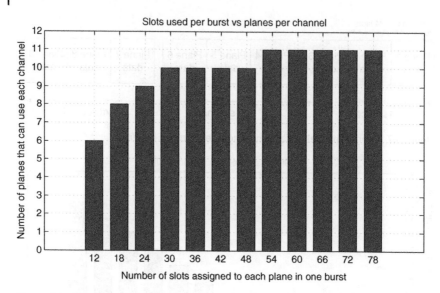

Figure 5.9 Number of planes that can occupy one channel for varying slots per burst.

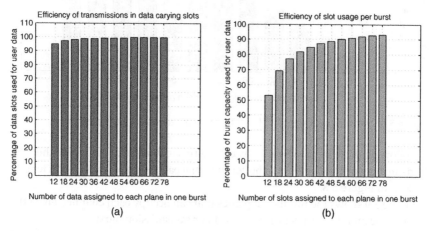

Figure 5.10 Efficiency of transmissions. (a) Data transfer efficiency in transmission slots. (b) Burst slot utilization efficiency.

Table 5.2 shows how the buffer length and slots per burst at each transmitting station affects the number of planes that can be supported using the 760 channels available in each cell. Without buffer and burst use, and having each plane transmit once every second, 760 of the 25 kHz channels would be able to support 4560 planes. Each of these channels can be used by six planes at a time. Using a two second buffer and burst scheme provides an improvement on the no buffer

Table 5.2 VDL mode 4 implementation comparison.

			VDL mode 4 results		
Slots/ burst	Buffer (s)	Planes/ channel	Total planes	Tx efficiency (%)	Slot efficiency
12	1	6	4560	94.64	52.91
18	2	8	6080	97.25	69.20
24	3	9	6840	98.18	77.12
30	4	10	7600	98.60	81.80
36	5	10	7600	98.88	84.89
42	6	10	7600	99.06	87.08
48	7	10	7600	99.20	88.72
54	8	11	8360	99.30	89.99
60	9	11	8360	99.38	91.00
66	10	11	8360	99.44	91.83
72	11	11	8360	99.49	92.51

scheme by increasing the number of required planes to 6080 by offering the ability for more aircraft to share each channel, while the scheme using an eight second buffer at each aircraft manages the load of 8360 aircraft.

The spectrum used for VDL communication is enough to supply 760 FDMA channels. Recall that these channels can be reproduced in a cellular-like frequency reuse fashion. With the aircraft in question operating across large areas, the number of channels offered by the system does not play a significant role. Each scheme discussed uses a very small percentage of the available channels.

5.4 Data Transfer – Internet Protocol Over VDL Transmission

This section uses the existing VDL mode 4 channel to transfer data as internet packets as shown in Figure 5.11. The wireless VHF channel will be the physical medium over which the internet packets are transmitted. This method of implementation, as with the satellite internet method, increases the complexity of the system. Unlike the satellite system channel configuration, we analyze the channels in VDL mode 4 as they are unamended. This is done because the time slots in VDL mode 4 can be concatenated to create one long burst, unlike the Iridium slots, implemented for multiple accessing purposes.

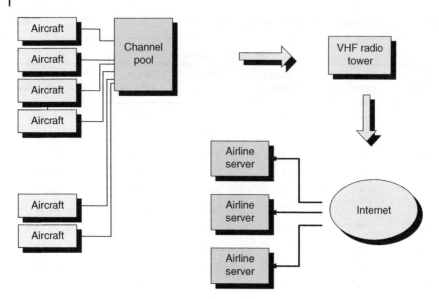

Figure 5.11 System model for Internet-based VDL implementation.

User Datagram Protocol (UDP) and Internet Protocol (IP) are the two protocols that will be used in the analysis, for the same reasons as those given for the satellite implementation, at the transport and data link layers.

5.4.1 Data Transfer with Internet Protocols

The VDL mode 4 channel is capable of bit rates of 19,200 bps. Each of the 25 kHz contains 4500 slots of length 13.33 ms. Using internet-based protocols, the RTS and CTS processes necessary to access the channel will be done away with. We do away with these system defaults, assuming that a more complicated algorithm will be implemented to deal with a controlled and exclusive channel accessing scheme. The packet structure used for the transmissions is similar to the 2298 bits long packets used in the satellite system evaluation.

5.4.1.1 Setup and Control

The segment assumes that the RTS and CTS transmissions will not be used with this implementation. We will use a channel accessing scheme that controls the number of planes using the channel (based on capacity requirements). The channel access scheme will monitor the number of stations that require a transmission period and assign each of the stations a designated number of slots for each burst. The VDL system does not have the built-in capability to deal with the transmission or reception of acknowledgments. We will therefore require that each transmission will need to reserve one slot for receiving an ACK from the receiver. The ACK is required because the UDP protocol used in the transport layer of the system does not provide a mechanism for acknowledging packet reception. As with the satellite implementation, UDP is only used between the aircraft and the radio tower receiver. The remaining leg of the transmission from the tower to the airline server may include Transmission Control Protocol (TCP) in sending the data upon entering the internet.

5.4.2 Packet and Frame Structure

The main difference between the structures of these frames and those used in the satellite system comes from the difference in the physical medium. The packets with 1536 user data bits are still encapsulated in the UDP and IP headers at the transport and data link layers. The UDP header of length 64 bits and the IP header of length 160 bits creates a packet of length 1760 bits. In adding physical layer overhead, we recall that the first slot used for data transfer in VDL mode 4 has an additional 16 bits of ramp-up power time, 24 bits of sync, 8 flag bits, 24 bits of source/destination address information and 9 bits of reservation information. Notice that even though the reservation of slots will not be used, information regarding the reservation type is still included in the framing to alert the system that the slots are prereserved. These overhead bits leave the first slot with 175 bits for the data packet. The last slot in every slot must contain 8 octets for forward error correction (FEC) and 24 bits for power ramp down. Therefore as was the case in the other VDL scenarios, the last slot will be able to ferry 168 bits of the data packet, while each in-between packet operates at 100% efficiency, carrying 256 bits. For a packet size of this length, we will be required to use eight time slots (6 slots carrying 256 bits, 1 slot carrying 175 bits, and one slot carrying 168 bits). These eight slots can carry a maximum of $(256 \times 6) + 175 + 168 = 1879$ bits. The efficiency of these transmissions is about 93.66% because of the excess data capacity that goes unused in each burst. An additional slot is reserved for the ACK, bringing the total number of slots used per burst to 9.

5.4.3 Data Packet Transmissions

With each burst occupying nine slots, and each slot being 13.333 ms long, we can see that each burst will take at least 120 ms. This means that in one second, each channel can maintain 8.333 bursts (1 s/120 ms). This means eight aircraft can use each channel to transmit internet packets over our VHF data link. These 8 bursts last 0.96 seconds, and leave 3 unused time slots in each second. Therefore a retransmission of a packet can be accomplished every three seconds (nine slots left over every three seconds) per channel.

With 760 channels available in each VDL cell, each cell will be able to facilitate the transfer of data for up to $8 \times 760 = 6080$ planes.

5.5 Number of Channels Needed to Support 5000 Planes

The big difference between the availability of channels in the satellite scheme and the VDL scheme is that the satellite channels available are once only over the entire region, while the VDL channels are available on a per cell average. With VHF transmission is limited to a propagation distance of about 200 miles maximum, we can assume each of our cells to be circular geometry that is approximately 400 miles in diameter to allow aircraft to use lower transmit power and have a good line of sight reception at the receiver. Each cell is assumed to be accommodated by at least one base station (Airport radio tower). These cells, therefore, have an area of about 125,664 m^2 each, and with the area of the united states equal to about 3,537,441 m^2, we claim that the entire area of the United States can be covered by about 28 (3,537,441/125,664) VDL cells. Another assumption we will make is that the air traffic over the area is uniformly distributed, meaning that one single cell will support about 180 planes (5000/28). The maximum number of aircraft that can use a single channel given the different buffer scenarios is 11. Therefore at least 17 channels of the available 760 channels per cell will be required to be dedicated to aircraft CVR/FDR data.

5.6 Expected Availability of Spectrum

The VHF spectrum reserved for aircraft has multiple functions. All of the frequency is used for communication, although it is primarily used for voice communication between aircraft crews and ground control facilities. A portion of the spectrum is also used to transfer different types of information about the flight. This communication is essential for the regular function of aviation. This same band of frequencies is also used by private and civilian aircraft. The number of used channels per cell could not be precisely determined, but we can assume that 90% of channels are used for these other applications. Total of 76 channels in every cell may be used or available for black box data transfer. With our best transmission schemes allowing 11 planes to access a single channel, one cell can be used for up to 836 aircraft, while the transmission scheme for one transmit per second, which allows the lowest number of planes to use each channel (6) but has no buffer time delay, can be used by 456 aircraft. With 28 cells across a land mass of area equal to that of the United States, the number of planes that

Table 5.3 VDL mode 4 implementation spectrum use comparison.

	VDL mode 4 channel requirement (for 5000 planes)		
Implementation	Planes/ channel	FDMA channels (per cell)	Spectrum usage (%)
54 slots per burst	11 (max)	17	$17/76 = 22.37$
36 slots per burst	10	18	$18/76 = 23.68$
24 slots per burst	9	20	$20/76 = 26.32$
18 slots per burst	8	23	$23/76 = 30.26$
12 slots per burst	6 (min)	30	$30/76 = 39.47$
IP based Transmissions	8	23	$23/76 = 30.26$

can transfer planes over the area is 12,768 when six planes use each channel, and 23,408 when a maximum of 11 planes share each channel. This is much more than the maximum air traffic density. Table 5.3 shows the percentage spectrum used for different VDL mode 4 implementations.

However, in real life, the distribution of air traffic is not uniform, and the density is higher around the airports, especially the bigger international airports. A cell that can support 836 aircraft at these locations will be useful if not necessary.

5.7 Summary

The VDL (VHF Data Link) mode 4 system has been demonstrated to be capable of handling aircraft data transfer effectively, provided that sufficient spectrum is allocated to VDL mode 4 aircraft data transfer. Each cell, with a diameter of 400 miles, can offer 760 FDMA channels from the entire airband frequencies. Research has shown that around 17 channels will be needed in each of these cells across the United States landmass for 5000 aircraft to transmit data. However, the number of channels required can be adjusted based on traffic density and demand in different areas. As the available frequency in the airband is primarily used for most aircraft communication, the channels used for other essential communications are not reused as they are in VDL mode 4.

The implementation of the VDL mode 4 system will require protocols that include an emergency transmission scene for critical situations where data can be streamed when needed. Additionally, proper planning and management of the airband frequencies are essential to avoid any interference with the VDL mode 4 data transfer. The system's success is highly dependent on the effective allocation of the required frequency spectrum and the implementation of appropriate protocols to handle different data transfer scenarios. However, if properly implemented, the VDL mode 4 system can significantly improve the efficiency and safety of aircraft data communication.

References

Joong-Won Bae, Hyoun-Kyoung Kim, In-Kyu Kim, Kwang-Jig Yang, and Tae-Sik Kim. Development of prototype VDL mode 2 system in Korea. In *2009 Integrated Communications, Navigation and Surveillance Conference*, pages 1–12, 2009. doi: 10.1109/ICNSURV.2009.5172842.

F. Bin Rahim and P. Breuer. Rohde & Schwarz aeronautical radio navigation measurement solutions, February 2011. URL https://www.rohde-schwarz.com/us/applications/aeronautical-radio-navigation-measurement-solutions-application-note-56280-15505.html.

S. Bretmersky, V.K. Konangi, and R.J. Kerczewski. Comparison of VDL modes in the aeronautical telecommunications network. In *Proceedings, IEEE Aerospace Conference*, volume 3, pages 3–3, 2002. doi: 10.1109/AERO.2002.1035250.

Xianchang Li, Kai Liu, and Jun Zhang. An improved long transmission protocol for VDL mode 4. In *2006 International Conference on Wireless Communications, Networking and Mobile Computing*, pages 1–4, 2006. doi: 10.1109/WiCOM.2006.371.

M. Mahmoud, C. Guerber, N. Larrieu, A. Pirovano, and J. Radzik. *Aeronautical Air–Ground Data Link Communications*. John Wiley & Sons, Inc., 2014.

S. Matale. Ground-based black box system implementation using satellite and VHF data link networks. Master's thesis, The University of Mississippi, Oxford, USA, 2010.

SAMYOG KC. High Frequency (HF), Very High Frequency (VHF), and Transponder usage in Aircraft, December 2020. URL https://www.aviationnepal.com/high-frequency-hf-very-high-frequency-vhf-and-transponder-usage-in-aircraft/.

Transport Canada. Transport Canada aeronautical information manual, October 2021. URL https://tc.canada.ca/sites/default/files/2021-09/AIM-2021-2-COM-E.pdf.

6

Cooperative Data Transmission Implementations

So far, procedures for transmitting data have been discussed using satellite communication links or VHF data Links. The satellite network is available on a global scale, but the availability of data channels is not guaranteed. The Iridium satellite system was originally intended for voice traffic for mobile services. Iridium is also used by the United States Military, and using these channels for voice calls (and other data transfer requirements) limits the number of available slots for aircraft data transfer. Using the spectrum of the Iridium system may also come at a cost, and minimizing operating costs is one of the biggest concerns to airlines. This fee may further limit the amount or number of channels that can be secured for use by aircraft for the purpose of black box data transfer. Additionally, the fact that VHF communication requires users to be in the range of a receiving station makes it extremely difficult for aircraft traveling long distances and over large areas without available receivers (like over oceans) to convey their information this way. To overcome the possible unavailability of channels or the inability to transmit data effectively using direct communication, it may become necessary to employ some form of transmission cooperation, not only between aircraft but between different systems. Using VDL mode 4 it is possible for aircraft that is out of range of a ground station to send data to other nearby aircraft and continue to hop from plane to plane until one plane can relay the information to a ground receiver. Obviously, data that has to be routed through aircraft will be received with a delay at the ground station. However, each plane can continue to send at least one burst of data per instance of time, based on the slot allocation technique used. Aircraft used for routing will transmit data on two separate channels, assuming that the antenna used for VDL mode 4 transmissions are capable of transmitting simultaneously in separate bands [Matale, 2010]. Besides the satellite and VDL mode 4 cooperative system, we will also discuss two concepts – aeronautical ad-hoc networks (AANET) and software-defined networking (SDN) – that can be used for airplane cooperative communications.

Real-Time Ground-Based Flight Data and Cockpit Voice Recorder: Implementation Scenarios and Feasibility Analysis, First Edition. Mustafa M. Matalgah and Mohammed Ali Alqodah.
© 2024 The Institute of Electrical and Electronics Engineers, Inc. Published 2024 by John Wiley & Sons, Inc.

6.1 VDL System-Based Relaying

In the VDL mode 4 data transfer implementation, not all aircraft can send information directly to a ground station. This is because VHF transmissions can only provide reliable communication up to a certain distance from the station. Because planes flying overseas are frequently not near a station where they can send data, we use VDL mode 4's ability to work without a ground control station and send data from one plane to another [Matale, 2010]. Figure 6.1 presents the idea of VHF relying implementation.

VDL mode 4 channels are plentiful at any given time (at least 76 channels for data relaying in a 400 miles diameter), due to the ability of the system to reuse the frequencies over and over, as long as co-channel interference is avoided. VDL mode 4 can be implemented without control from a ground base station and system control can be carried out by the in-flight aircraft. For the system model, it will initially be assumed that each aircraft out of range of a VHF ground station, will be able to connect with at least one aircraft within the VHF ground station range. With this assumption, we say that all the data routed will have at most one hop. The aircraft used to do the routing will be capable of utilizing two burst frames on separate channels at any given time, as long as the necessary channel slot reservations are made. Therefore, both aircraft in the path will be able to continually send data as before, with at least one burst per predetermined time. We also work with the assumption that the channels used by each of the VHF radio tower range aircraft to send data to the relaying aircraft are not the same channels used by the

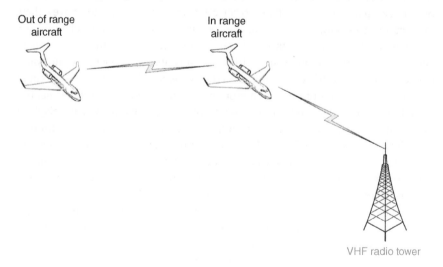

Out of range
aircraft

In range
aircraft

VHF radio tower

Figure 6.1 VHF relay implementation.

in-range aircraft to relay data to the radio tower. This means that the channels used by the relaying aircraft are not being shared by out-of-range aircraft. The data arriving at the ground VHF station from the out-of-range aircraft will be delayed by an additional time frame, based again, on the slot allocation in use. Upon entry into the range of a ground station, the aircraft will then resume normal burst transmissions to the ground, after having secured a reservation for an available burst frame time in any of the available channels.

Air-to-air communication in VDL mode 4 is implemented in the same way as air-to-ground communication. An aircraft that intends to begin transferring data to another aircraft must first reserve a slot (or slots) for transmission by sending at least three Request to Send (RTS) messages within each available channel. In the air-to-air scheme, the aircraft that will relay the data confirms the request and sends a CTS message to the transmitting station, informing it of the reserved time slots. The first plane can then send its data to the relay plane. The relay plane will then reserve a transmission on another channel to forward the first aircraft data.

The required number of bits originating at each aircraft is still 1536. In the no buffer and burst scheme discussed previously in Chapters 4 and 5, each burst from each aircraft needs to have a minimum of seven data carrying time slots (5 slots carrying 256 bits, the first slot carrying 175 bits, last slot carrying 168 bits). With each burst requiring the additional RTS (three slots), CTS (one slot), and acknowledgement (ACK) (one slot), the total number of slots for each burst becomes 12, for each aircraft. As was the case in the direct transmit mode, we say that six aircraft will be able to utilize one 25 kHz channel, except that, these six aircraft are six that are out of range. The in-range aircraft will need two bursts at a time, one for their own data and one for the data they are relaying. In VDL mode 4, the larger chunks of data can be transmitted more effectively because of the additional overhead and signaling required for each transmission. However, with data intended for two different destinations, one stream of data, with one overhead block, cannot be used. This means that the relaying aircraft will need to send two separate bursts, as opposed to a more efficient single but longer burst. Therefore, three aircraft can use one 25 kHz channel in the relaying phase (a different channel pool). There must be twice as many channels available to the in-range aircraft as the out-of-range aircraft. Ideally, with 76 channels available in each pool, and each channel capable of accommodating six out-of-range aircraft, one cell can have a maximum of $76 \times 6 = 456$ out-of-range planes. With one channel accommodating three planes in the in-range phase, $76 \times 3 = 228$ planes can occupy one cell. This means that more than one cell would need to be used in the in-range phase of the transmission, which is not an inconceivable or unrealistic scenario. However, we will use an exclusive channel assignment technique for both in-range and out-of-range cells. We use this to help with co-channel interference and allow one-on-one cell interaction [Matale, 2010]. Total of 76 flight data transfer channels are available inside each pool (cell). We have assumed that each cell has a diameter of 400 miles. Using the eastern coast of the United States as an example, approximately 2069 miles long, we show how relaying data can be accomplished in VDL mode 4 transmission. To minimize the delay time of received data at the ground station, only the one burst per second method will be used in the relaying application. In this method, every plane transmits data at least once a second. The data from the out-of-range aircraft will therefore be about two seconds delayed. Along this coastline, we will say that there are six cells for out-of-range planes and six cells in the range. It is necessary that the in-range aircraft can use channels in both in-range and out-of-range cells. For this to happen, the channel frequencies will have to be isolated by a large enough spectrum to avoid co-channel interference. With 76 channels available, this can be accomplished by using a predetermined number of these channels for the relaying stage and another set of channels for the out-of-range transmissions. Twice as many transmissions are required in the relaying phase compared to the out-of-range transmission phase. Twice as many channels are needed in the in-phase transmission stage, used for relaying the data from out-of-range planes and the data from the relaying plane itself. As a channel assignment technique, 25 channels can be used in the out-of-range cell pool (as depicted in Figure 6.2), and 51 can be used in

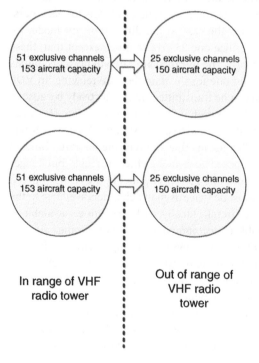

Figure 6.2 Channel sharing in cell pairs.

51 exclusive channels
153 aircraft capacity

25 exclusive channels
150 aircraft capacity

51 exclusive channels
153 aircraft capacity

25 exclusive channels
150 aircraft capacity

In range of VHF radio tower

Out of range of VHF radio tower

the in-range cell pool. Each of the 25 channels that are used outside of the range of VHF radio towers, using the one burst per second scheme, can be utilized by six aircraft. The total number of planes that can occupy one of these 400 miles diameter cells is therefore (25 channels × 6 planes per channel = 150). The number of planes that can utilize each channel in the cell that has the capability to transmit to a VHF radio tower is three, because each plane in these cells will be required to send information for two planes. The total or maximum number of planes in this cell will therefore be (51 channels × 3 planes per channel = 153). This is one channel assignment that can be used. We use this particular one to have the same number of planes in each transmission stage and maintain a one-to-one relationship. Referring back to the example of the United States eastern seaboard of approximate length 2069 miles, about six cells can be used for each of the transmission phases (in range and out of range). Altogether these cells can supply enough transmission capacity for 150 × 6 = 900 planes [Matale, 2010].

When different traffic intensities are encountered, these channel assignments can be reconfigured to accommodate the specific traffic distributions. Trans Oceanic flights are not ideally within one hop of a VHF radio tower that can be used to connect it to its storage server on the ground. When aircraft are out in the middle of the ocean, more than one hop will be necessary to complete the transfer. With each hop in the network, the number of planes that can use each channel is reduced by at least half. This is because any aircraft relaying data will be responsible for at least two transmissions (more, if relaying for more than one plane). Using the one burst per second scheme that allows six transmissions per second on each channel, a maximum of three planes can use one channel to relay data for one plane each. Alternatively, one plane can use a single channel to transfer data for five aircraft and itself and any other combination of plane data that fits into the six bursts per channel window. There are certain circumstances in which it is essential to have collaboration even between the two systems, the satellite, and the VDL system, this is what will be discussed in Section 6.2.

6.2 VHF and Satellite System Cooperation

Though we have shown that each system alone can handle large quantities of air traffic, the availability of the capacity is made on assumptions. In the instance that the necessary resources are unobtainable or inadequate (due to factors such as cost), there may be a need for airliners to use a combination of both of the technologies discussed in this chapter. In other words, an alternative method for transmitting black box data to the ground is to use the Iridium satellite system and VDL mode 4 together. Aircraft will be required to use either of the technologies based on the phase of flight the plane is in. A basic implementation of this scheme will mandate that flights below a certain altitude use VDL transmissions for data transfer, while flights at higher altitudes use satellite technology. Alternatively, the flight length can be used to determine which aircraft use which scheme. Long haul flights or flights that last longer than five hours, for example, will be assigned Iridium satellite channels, while shorter duration flights can use the VDL system. Using the flight length may have an advantage over the altitude-based separation because the altitude separation scheme would require aircraft to be equipped with the necessary hardware for both transmissions and may be costly. Handing off the communication from the satellite system to the VDL system may also be a little more complex. The alternative method of dividing planes by flight duration may be more cost- and complexity-efficient. The operating airlines usually use particular aircraft for specific flight durations, meaning that those planes would need the equipment for only that one type of transmission, transmission that will be the same for the entire flight, removing the need to handoff from system to system [Matale, 2010]. Following is a discussion and summary of AANET, a technology with promising applications in cooperative communication.

6.3 Aeronautical Ad-hoc Network (AANET)

To transfer the Aircraft FDR/CVR data to the ground station via a network of neighboring aircraft, we address AANET, as shown in Figure 6.3, as a good choice for this type of comparative communication.

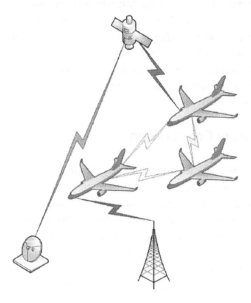

Figure 6.3   AANE. Source: Bilen et al. [2022]/Institute of Electrical and Electronics Engineers/CCBY 4.0.

Aeronautical Ad-hoc Networks, or AANET for short, have been suggested as an innovative and potentially effective solution to the challenges given by satellite and air-to-ground networks. These networks work to facilitate connections between aircraft. Over the AANET, data packets are passed from one aircraft to another, and the receiving aircraft is the one responsible for ensuring that they are successfully delivered to the location to which they were originally addressed. The AANET is distinguished from other flying Ad-Hoc networks (FANETs) that operate in the world of vehicle networks by virtue of its highly customizable design. Without a hub or controlling entity, AANETs are formed when airplanes in the sky establish direct communications with one another. These connections allow data to travel from an origin aircraft to the destination. Using the VHF band's 119–137 MHz spectrum, the same as VDL links, these links typically exhibit line of sight (LOS) characteristics, making them have a reasonably high signal-to-noise ratio. Linking airplanes in the air is often carried out according to the distance at which radio signals may travel from one place to another. Aircraft can communicate with one another via omnidirectional transmission if the distance between them is less than the transmission range. Every plane in the AANET network functions as a router during the time that packets are being routed across the air-to-air links. Also, the plane can connect to the Internet through satellite or A2G connectivity. As a result, the AANET framework combines the benefits of both satellite connectivity and the A2G network [Bilen et al., 2022].

In the AANETs, there are three distinct layers. The satellite layer, the airplane layer, and the ground layer make up the top, middle, and bottom of this tiered structure, respectively. Using interlayer links, each layer might communicate with the others. Through satellite-to-air and satellite-to-ground connectivity, the satellite layer can communicate with the air and ground layers. Aircraft can communicate with the A2G ground stations via air-to-ground communications. The aircraft layer also has air-to-air links formed between the planes to make an AANET [Bilen et al., 2022]. Figure 6.4 shows the AANET different layers.

AANET channels, as previously discussed, exhibit LOS, which can be modeled as free space loss. However, the propagation effects must be considered in the design of these LOS systems. The authors of Bilen et al. [2022] also investigate the attenuations caused by gases and rain. They believe oxygen absorption, rain, and cloud attenuation can be incorporated into the AANETs' free space path loss model. Because of the long range of aeronautical networks, oxygen absorption must also be considered. Rainfall and atmospheric gaseous cause absorption and scattering at frequencies above 5 GHz. This situation causes increased channel error rates, which increases transmission losses. In particular, the air-to-air links in AANET were easily disturbed due to the rain attenuation effect. This scenario, like the mobility effect case, results in a rapid topology change. Therefore, the propagation model must consider rain attenuation for frequencies above 5 GHz.

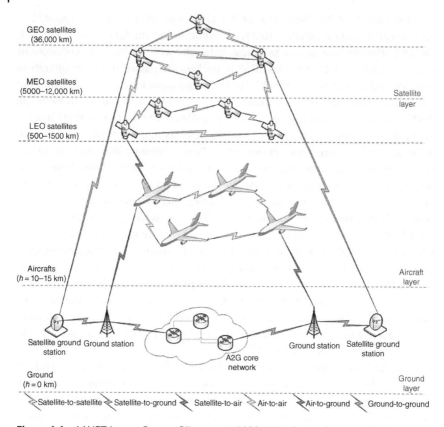

Figure 6.4 AANET layers. Source: Bilen et al. [2022]/IEEE/Public Domain CC BY 4.0.

Mobile ad-hoc networks are proposed for use in oceanic flight routes in Tu and Shimamoto [2009], with the idea being that any aircraft's "location report" can be relayed to the appropriate ground station by route of other aircraft in the local ad-hoc network as depicted in Figure 6.5. Assume packets are permitted to remain at each aircraft while waiting for link availability. In that situation, packets are permitted to remain at each aircraft while waiting for link availability; therefore, each aircraft must be equipped with one router to route packets to the next destination and one server to store these packets for a predetermined delay time. Each aircraft first listens to the information broadcast by its neighbors during any frame period before building a routing table with distance-based and load-based priority. Each plane's routing table must contain the most fundamental data, including its location, neighboring nodes distances, data loads, and relative orientation. The authors of Bilen et al. [2022] suggest the following packet relaying algorithms.

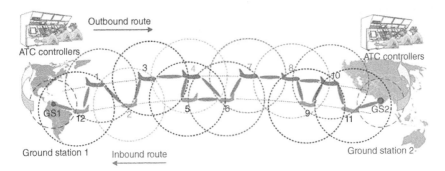

Figure 6.5 Model for oceanic flights route employing mobile ad-hoc network. Source: Adapted from Bilen et al. [2022].

1. *Algorithm 1*: Each aircraft relay packets to ahead, furthermost, and only the same direction aircraft.
2. *Algorithm 2*: Each aircraft relay packets to the ahead and furthermost aircraft. However, if there is no same-direction aircraft, it can select opposite-direction aircraft.
3. *Algorithm 3*: The whole airspace is divided equally into several parts where each part is assigned one ground station. If an aircraft belongs to some part, its packets will be relayed to the next aircraft in the same part which is closest to the ground station of that part. If there are no same-direction aircraft, it selects opposite-direction aircraft.

According to simulation results from Tu and Shimamoto [2009], this technique helps the aeronautical authority (Air Traffic Control (ATC) center) to safely decrease the delay in time or horizontal separation in the distance between consecutive flights. As a result, it is possible to increase the number of aircraft flying on maritime routes.

6.4 Software-Defined Networking

SDN is another research area that can aid in transmitting FDR/CVR data recorders in cooperative airplane communications. The dynamic network architecture made possible by SDN allows for transforming static network backbones into feature-rich service delivery platforms. OpenFlow-based SDN design separates the network's control and data planes, making it easier to program and maintain at scale, just like computer infrastructure. SDN was first used to describe the research and development efforts surrounding OpenFlow at Stanford University [Greene, 2009]. The initial definition of SDN focused on a network architecture in which the data plane's forwarding state was governed by a separate, remote control plane. The networking industry has strayed far from this original definition of SDN, now commonly referring to anything that requires software as being "SDN" [Kreutz et al., 2015]. The promise of overcoming the limitations of existing network infrastructures is at the core of the SDN networking paradigm. It separates the network's control logic (the control plane) from the actual routers and switches forward traffic (the data plane). Network switches are reduced to essential forwarding devices by separating the control and data planes. In contrast, the control logic is implemented in a logically centralized controller (or network operating system), facilitating policy enforcement and network reconfiguration and evolution. Figure 6.6 depicts the SDN Architecture as illustrated in [Kreutz et al., 2015].

In Wang et al. [2019], using the air-to-air aircraft radio links, the authors propose a software-defined wireless flight recorder (SD-WFR) architecture to transmit flight data in real time. Satellite links can also be used as a backup plan, only when air-to-air links are unavailable. Authors of Wang et al. [2019] developed a time-expanded connectivity graph model illustrated in Figure 6.7, based on the physical position and predictable trajectory of each commercial aircraft. To characterize the time-varying sporadic but predictable communication opportunities among aircraft, satellites, and ground base stations. The authors formulate a Mixed integer linear programming (MILP) problem based on this graph model to minimize the amount of data sent via satellites and use a branch and price algorithm to find the optimal flow configurations efficiently. In the event of unexpected network changes, the centralized controller must recalculate the optimal flow configuration and decide whether or not to apply it to the network immediately. They propose a reroute policy to keep the gap as small as possible while minimizing the cost of flow.

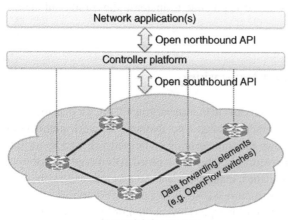

Figure 6.6 SDN architecture in Kreutz et al. [2015]. Source: Kreutz et al. [2015]/with permission of IEEE.

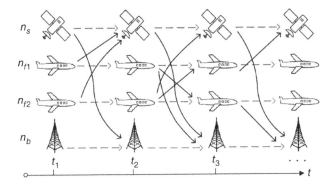

Figure 6.7 Time-expanded connectivity graph in Wang et al. [2019]. Source: Wang et al. [2019]/with permission of IEEE.

The proposed SD-WFR over the North Atlantic consists of three components: the ground system, the aerial system, and the satellite system.

1. *Ground system*: The ground system consists of A2G base stations and a ground control station managed by a centralized controller. The centralized controller has information about the entire network, including aircraft position, satellite position, link capacity, and latency Using this data, it determines the most efficient flow topologies for transmitting data from each plane.
2. *Aerial system*: The aerial system comprises all commercial aircraft flying over the North Atlantic.
3. *Satellite system*: The satellite system includes all the available satellites served for aeronautical communication over the North Atlantic. When aircraft flies over the ocean where no base station has been deployed, and there is no other aircraft in the line of sight, it would have to resort to satellite transmission. Satellite communications provide worldwide coverage and relatively high-speed links but at the expense of a high cost. Therefore, the satellites are only used to transmit the flow configuration and recorder data when aircraft meet the above difficulties.

Numerical experiments were conducted, and the effectiveness of several reroute rules in realistic aeronautical scenarios was assessed to ensure the proposed SD-WFR architecture is feasible and efficient.

6.5 Summary

In this chapter, we explored the implementation of cooperative data transmission methods in aviation to transmit the FDR/CVR data. The chapter begins with discussing the limitations of existing data transmission methods, such as satellite communication links and VHF data links. The Iridium satellite system, initially designed for voice traffic, poses limitations on the availability and number of channels for aircraft data transfer. VHF communication requires proximity to a receiving station, making it challenging for long-distance flights over areas without coverage. To overcome these challenges, the chapter introduces the concept of transmission cooperation, both between aircraft and different systems. It explores the use of VDL mode 4, which allows aircraft out of range of a ground station to send data to nearby aircraft, creating a relay network until the information reaches a ground receiver. This method incurs a delay but enables continuous data transmission using burst frames and slot allocation techniques.

The chapter also discusses two additional concepts for cooperative communication: AANET and SDN. AANETs enable direct connections between aircraft in the air, utilizing the VHF band and LOS characteristics. Each aircraft acts as a router, facilitating the transmission of data packets across the network. SDN, on the other hand, focuses on centralized control and programmability of network infrastructure, allowing for more flexible and efficient data transmission in a cooperative setting. Furthermore, the chapter explores the cooperation between VHF and satellite systems. It suggests using a combination of Iridium satellite system and VDL mode 4 based on flight altitude or duration to optimize data transmission. Flights below a certain altitude or shorter durations can utilize VDL mode 4, while higher altitude or longer flights can rely on satellite technology.

References

Tuğçe Bilen, Hamed Ahmadi, Berk Canberk, and Trung Q. Duong. Aeronautical networks for in-flight connectivity: A tutorial of the state-of-the-art and survey of research challenges. *IEEE Access*, 10:20053–20079, 2022. doi: 10.1109/ACCESS.2022.3151658.

K. Greene. *TR10: Software-Defined Networking*. Cambridge: Technology Review, Inc., February 2009. Retrieved from https://www.proquest.com/other-sources/tr10-software-defined-networking/docview/2820784065/se-2.

Diego Kreutz, Fernando M. V. Ramos, Paulo Esteves Veríssimo, Christian Esteve Rothenberg, Siamak Azodolmolky, and Steve Uhlig. Software-defined networking: A comprehensive survey. *Proceedings of the IEEE*, 103(1):14–76, 2015. doi: 10.1109/JPROC.2014.2371999.

S. Matale. Ground-based black box system implementation using satellite and VHF data link networks. Master's thesis, The University of Mississippi, Oxford, USA, 2010.

Ho Dac Tu and Shigeru Shimamoto. Mobile ad-hoc network based relaying data system for oceanic flight routes in aeronautical communications. *International Journal of Computer Networks and Communications (IJCNC)*, 1(1):33–44, 2009.

Yuanyuan Wang, Kai Wan, Chi Zhang, Xia Zhang, and Miao Pan. Optimized real-time flight data streaming via air-to-air links for civil aviation. In *ICC 2019 - 2019 IEEE International Conference on Communications (ICC)*, pages 1–6, 2019. doi: 10.1109/ICC.2019.8761955.

7

UAV Wireless Networks and Recorders

Unmanned aerial vehicles (UAVs) have drawn increased attention from regulators due to a number of their benefits, including their adaptability, agility, ease of installation, and comparatively low operating costs. These elements have stoked interest in the potential military and civilian applications of UAVs. UAVs are predicted to play a major role in the creation of future wireless networks, supporting high-rate transmissions and enabling wireless broadcasting. UAVs offer unique advantages like flexible deployment, strong line-of-sight connection linkages, and greater design freedom with controlled mobility, in contrast to typical fixed infrastructure communications. UAVs can therefore deliver dependable and affordable wireless communications from airborne platforms. This chapter aims to delve into two major concerns regarding UAVs. First, it will explore the role UAVs can play in deploying real-time recorders on airplanes. The use of UAVs in this context has the potential to enhance the monitoring and recording capabilities of aircraft during flight. Second, the chapter will discuss the possibility of UAVs themselves requiring flight data recorder (FDR) recorders for safe operation. This consideration highlights the importance of implementing appropriate recording systems on UAVs to ensure their operational safety and compliance with regulatory requirements.

7.1 UAV Communication Networks

Wireless networks are divided into two categories: those that rely on a preexisting infrastructure, known as infrastructure-based networks (IBN), and those that do not, known as infrastructure-less networks (ILN) or ad-hoc networks. wireless sensor networks (WSNs), wireless mesh networks (WMNs), and mobile ad-hoc networks (MANETs) are all examples of ILNs. The MANETs are divided into two subgroups: Vehicular ad-hoc networks (VANET) and flying ad-hoc networks

Real-Time Ground-Based Flight Data and Cockpit Voice Recorder: Implementation Scenarios and Feasibility Analysis, First Edition. Mustafa M. Matalgah and Mohammed Ali Alqodah.

(FANET), unmanned aerial vehicle communication networks (UAVCN), a subgroup of the FANETs are of interest in this chapter.

ILN have an advantage in many situations where an IBN cannot meet the requirements. To accomplish their goals, some apps require a temporary network setup that can be completed quickly that's why a network that does not rely on physical infrastructure is needed to meet the specific requirements. Compared to conventional network paradigms, the characteristics of the ILN are distinct. Because of the absence of infrastructure, the nodes communicate without fixed connectivity. The topology for these nodes is dynamic and continually evolving [Nawaz et al., 2021]. Due to their importance as a form of infrastructure-free network, MANETs have received a lot of attention. This is a crucial aspect of a wireless network environment because it allows mobile nodes to communicate at any time and in any direction. Auto-configuration is a key component of MANET technology, which allows nodes to join or leave the network during communication. Communication between nodes in a MANET can be autonomously arranged and set up [Nawaz et al., 2021].

FANET is a subset of MANET that has its unique characteristics (Figure 7.1). There are, however, several key distinctions between FANET and other ad-hoc networks, such as the significantly higher mobility degree of FANET nodes compared to that of MANET or VANET nodes. In contrast to the ground-based nodes used in MANETs and VANETs, which are people and vehicles, respectively, the nodes used in FANETs are flying. The network topology of a FANET is shown in Figure 7.2 can vary more frequently than a regular MANET or even a VANET due to the high mobility of FANET nodes [Oubbati et al., 2017, Nawaz et al., 2021]. The existing ad-hoc networks seek to establish peer-to-peer connections. Furthermore, peer-to-peer links are required in FANET for UAVs to coordinate and collaborate effectively. Also, similar to wireless sensor networks, it primarily functions as a data collector, taking readings from its surrounding area and sending them to a centralized location for analysis. For this reason, FANET needs to accommodate peer-to-peer communication and converge cast traffic. In contrast to MANETs and VANETs, the typical distance between FANET nodes is substantially greater. Therefore communication range between UAVs must also be greater than that of MANETs and VANETs. This phenomenon has implications for radio connections, hardware circuits, and the physical layer. Sensors in multi-UAV systems may be of varying types, with correspondingly unique requirements for how and when that data is delivered [Bekmezci et al., 2013, Oubbati et al., 2017, Nawaz et al., 2021].

Suppose a multi-UAV communication network is infrastructure based. In that case, the operation area will be limited to the communication coverage of the infrastructure, and a UAV will not be capable of operating if it cannot connect with the infrastructure. However, FANET can expand its coverage since it relies on UAV-to-UAV data to link UAVs to infrastructure ones. If a FANET node loses

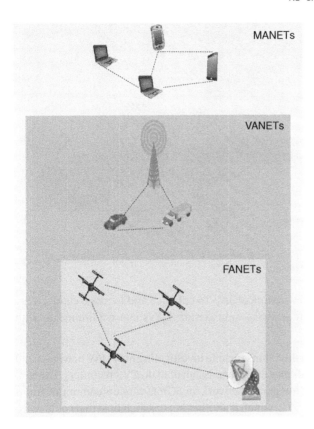

Figure 7.1 FANETs subclass. Source: Oubbati et al. [2017].

contact with the network's base, it can continue functioning by exchanging data with other UAVs [Bekmezci et al., 2013]. Multi-hop communications among aerial nodes like UAVs, aircraft, and helicopters are necessary for all FANET applications. Recent technological advances have enabled the mass distribution of UAVs equipped with a wide range of sensors and computing hardware. As a result, UAVs are best suited for tasks that need collaboration and more equal task distribution via multi-hop wireless communications [Nawaz et al., 2021].

The FANET architecture can be described briefly as follow. FANETs adhere to a norm analogous to that of MANETs, in which the dynamic nature of the network is a direct result of its mobile nodes. A conventional FANET's nodes can communicate in two ways [Nawaz et al., 2021]:

1. *Air-to-air wireless communications*: To bypass the transmission range constraints caused by communication between UAVs and ground base stations, UAVs can communicate with each other utilizing a pure ad-hoc architecture.

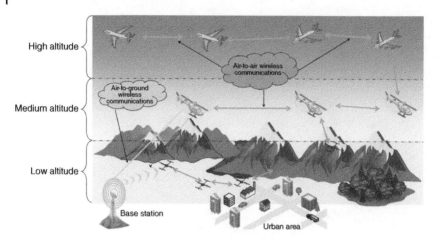

High altitude

Air-to-air wireless communications

Air-to-ground wireless communications

Medium altitude

Low altitude

Base station

Urban area

Figure 7.2 Wireless communications between FANET nodes. Source: Oubbati et al. [2017]/with permission of Elsevier.

2. *Air-to-ground wireless communications*: To enhance and increase connectivity and provide additional services, only certain UAVs can communicate with infrastructures.

Possible future research directions include investigating how UAV networks and FANET, in general, can enhance the transmission of FDR/CVR recorders. The concept of a space-air-ground integrated network for 5G/6G wireless communications will be discussed in Section 7.2, which is another hot research field that could assist the transmission of FDR/CVR recorders.

7.2 Space-Air-Ground Integrated Network for 5G/B5G Wireless Communications

The space-air-ground integrated 5G/6G wireless networks, which could enable wireless broadcast and handle high-rate transmissions, are likely to include UAVs as a key component. The authors of Li et al. [2019] presented a three-layer cooperative network design integrating the air, ground, and space layers to allow 5G/B5G wireless communications as illustrated in Figure 7.3.

The space-based network includes satellites or constellations in various orbits, ground stations, and network operations control centers. Inter-satellite links can create a worldwide space-based network using multicast and broadcast techniques. Meanwhile, interconnections are established with neighboring satellites and ground cellular networks via satellite-to-UAV and satellite-to-ground

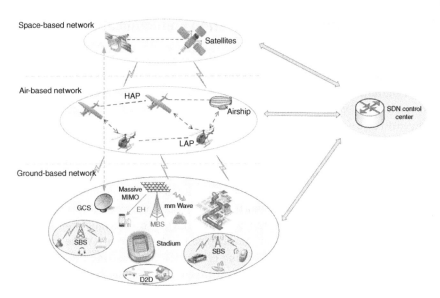

Figure 7.3 Illustration of space-air-ground integrated networks. Source: Li et al. [2019].

connections. The space-based network has the potential to provide global coverage on Earth with services including search and rescue, navigation, earth observation, and communication and relaying. Satellite-to-UAV communication is a critical component in developing an integrated space-air-ground network. The satellite-to-UAV channel relies mostly on the LoS link and suffers from rain attenuation in the Ka-band and above. UAVs can communicate with satellites in various orbits according to their applications and equipment. Since its location relative to Earth remains constant, geosynchronous satellites are employed for satellite-to-UAV communication. The air-based network described in Li et al. [2019] is analogous to the FANET concept outlined in Section 7.1. The ground-based network's mobile users, including those with cell phones, autonomous vehicles, IoT devices, and so on, are serviced by a heterogeneous radio access network consisting of both macro cells and small cells. In this way, 5G wireless networks and disruptive technologies can coexist. The millimeter wave (mmWave) band, energy harvesting, non-orthogonal multiple access (NOMA) transmission, and direct-to-direct communication are all promising prospects for 5G and beyond cellular networks. Li et al. [2019] addressed physical layer candidates to enhance the system performance of UAV communication in 5G networks including millimeter wave communication, NOMA transmission, cooperative radiation (CR), and energy harvesting.

7.3 Integrating UAVs Into Aviation Communication

The soaring demand for high-speed wireless access, driven by the widespread use of mobile devices and emerging technologies like drones and IoT gadgets, has placed significant strain on existing wireless networks. To meet this challenge, the emergence of 5G cellular systems and beyond seeks to provide ubiquitous connectivity to diverse wireless devices. Within this framework, UAVs are poised to play a pivotal role, offering reliable and high-speed wireless connectivity not only to stationary users but also to people in transportation networks. Unlike 4G networks, which have limitations in connecting various wireless devices, 5G and beyond are designed to accommodate a wide range of requirements. The proliferation of IoT devices has resulted in increased data traffic, surpassing the capabilities of existing terrestrial wireless systems operating below 6 GHz. To overcome this hurdle, mmWave communications exploit unoccupied bandwidth at mmWave frequencies, proving instrumental in alleviating congestion and meeting the demands of 5G networks. UAVs can leverage mmWave communications to assist existing wireless networks, especially in future 5G applications [Ghamari et al., 2022].

Successful testing of aerial base stations operating at mmWave frequencies has demonstrated their stability and reliability in achieving multi-gigabit-per-second data rates. UAVs can act as aerial base stations, acting as a complement to existing cellular networks, particularly during periods of downlink traffic overload. Advanced techniques such as weighted expectation maximization algorithms and contract theory can optimize user distribution and downlink traffic demand in congested areas, ensuring efficient resource allocation. Looking ahead, UAV-to-UAV and satellite-to-UAV communications represent future research directions. Establishing swarm formations of UAVs to create multi-hop wireless networks is vital for providing wireless communication services over vast geographical areas. However, frequent interruptions in radio communication links between nearby UAVs, resulting from their rapid mobility, pose challenges for conventional routing protocols. Innovative control mechanisms are needed to ensure acceptable services, including collision avoidance to guarantee the safe operation of UAVs. Furthermore, the development of propagation models for satellite-to-UAV communication is in its early stages and will be a subject of future research, offering insights into the unique characteristics of this cutting-edge communication channel [Ghamari et al., 2022].

7.4 UAV Recorders

The use of UAVs in densely populated areas, while ensuring human safety and avoiding property damage, is becoming more widely anticipated. However, accidents involving UAVs are still a possibility and unfortunately, their occurrence is on the rise. Traditionally, FDRs installed on aircraft have been utilized for accident investigation, but this method relies on recovering the drone itself. Furthermore, physical FDRs are not suitable for conducting forensic digital analysis of UAV flights because they are too heavy to be installed on lightweight drones. The design of data recorders for UAVs presents several challenges. First, these recorders need to rapidly acquire and store a large volume of data due to the high sampling rate of status information. Additionally, the available space for installing a data recorder on a UAV is extremely limited, considering the restricted payload capacity of these aircraft. Therefore, it is crucial for the data recorder to be lightweight and compact in size. Moreover, in order to withstand the intense forces experienced during launch or in the event of an impact, the recorder must be resistant to impact, vibration, and crashes [Liu et al., 2011]. Given the aforementioned limitations of conventional UAV FDRs, there is a growing interest in exploring wireless UAV FDRs as a promising area for research. These wireless FDRs offer potential solutions to the challenges posed by traditional recorders.

7.5 Summary

In this chapter, the role of FANETs, specifically UAVs, in wireless communication networks is explored. UAVs have garnered significant attention due to their adaptability, agility, and cost-effectiveness, leading to their application in various military and civilian contexts. The chapter focuses on two key aspects: the deployment of real-time recorders on airplanes using UAVs and the potential need for FDR recorders on UAVs themselves. The chapter commences by discussing the advantages of UAVs in wireless communication networks. Unlike IBN, UAVs offer flexible deployment, strong line-of-sight connections, and greater design freedom with controlled mobility. They enable reliable and affordable wireless communications from airborne platforms. The concept of FANETs, which are a subset of MANETs, is introduced to emphasize the unique characteristics and communication requirements of UAVs. Section 7.2 introduces the concept of a space-air-ground integrated network for 5G/6G wireless communications. This network integrates satellite-based, air-based (FANETs), and ground-based networks to provide global coverage and support high-rate transmissions. The utilization of UAVs as aerial base stations is examined, as they can alleviate congestion and optimize resource allocation in cellular networks. Future research directions, such as UAV-to-UAV and satellite-to-UAV communications, are discussed, shedding light on the challenges and opportunities involved in establishing multi-hop wireless networks and developing propagation models for satellite-to-UAV communication. Lastly, the chapter underscores the significance of UAV recorders for accident investigation and flight data analysis. Despite efforts to ensure UAVs fly safely in populated areas without posing risks to human safety or property, accidents can still happen. Traditional FDRs employed in aircraft are not suitable for lightweight drones, necessitating the exploration of wireless UAV FDRs. The design challenges associated with UAV recorders, including data acquisition, limited space, and durability, are elucidated. Wireless FDRs are presented as a potential solution to overcome these limitations.

References

İlker Bekmezci, Ozgur Koray Sahingoz, and Şamil Temel. Flying ad-hoc networks (FANETs): A survey. *Ad Hoc Networks*, 11(3):1254–1270, 2013. ISSN 1570-8705. doi: 10.1016/j.adhoc.2012.12.004.

Mohammad Ghamari, Pablo Rangel, Mehrube Mehrubeoglu, Girma S. Tewolde, and R. Simon Sherratt. Unmanned aerial vehicle communications for civil applications: A review. *IEEE Access*, 10:102492–102531, 2022. doi: 10.1109/ACCESS.2022. 3208571.

Bin Li, Zesong Fei, and Yan Zhang. UAV communications for 5G and beyond: Recent advances and future trends. *IEEE Internet of Things Journal*, 6(2):2241–2263, 2019. doi: 10.1109/JIOT.2018.2887086.

Guanghui Liu, Jun Zhou, and Xiaozhou Yu. The design and realization of small-sized UAV solid data recorder. In *Proceedings of the 2011 International Conference on Electronic & Mechanical Engineering and Information Technology*, volume 6, pages 3086–3091, 2011. doi: 10.1109/EMEIT.2011.6023740.

Haque Nawaz, Husnain Mansoor Ali, and Asif Ali Laghari. UAV communication networks issues: A review. *Archives of Computational Methods in Engineering*, 28:1349–1369, 2021.

Omar Oubbati, Abderrahmane Lakas, Fen Zhou, Mesut Güneş, and Mohamed Yagoubi. A survey on position-based routing protocols for Flying Ad hoc Networks (FANETs). *Vehicular Communications*, 10:29–56, 2017. ISSN 2214-2096. doi: 10.1016/j.vehcom.2017.10.003.

8

Future Aviation Communication

This chapter will primarily address advancements in aviation communications research that offer potential improvements for Flight Data Recorder (FDR)/ Cockpit Voice Recorder (CVR) data streaming. The initial focus will be on introducing system wide information management (SWIM), a platform developed by International Civil Aviation Organization (ICAO) to facilitate the sharing of aeronautical data. Subsequently, recent progress in air-to-ground (A2G) and air-to-air (A2A) communication will be explored, followed by a discussion on leveraging machine learning to enhance future communications.

8.1 System Wide Information Management (SWIM)

The industrial sectors have shown significant interest in leveraging big data analytics, cloud computing, and the Internet of Things (IoT) to develop innovative services. However, the integration of air traffic management (ATM) information systems faces challenges due to diverse data formats and exchange protocols. Nonetheless, there is a growing enthusiasm for utilizing artificial intelligence (AI) and big data analytics to make informed decisions based on aeronautical information shared through IoT and the cloud. To tackle these challenges, ICAO has standardized a future platform called SWIM, which facilitates the sharing of aeronautical big data, including flight, weather, and surveillance data, by implementing standardized formats and protocols. This ensures seamless system integration [Morioka et al., 2020]. In response to the limitations of the current aeronautical mobile communication system in adopting SWIM, ICAO has introduced the Aeronautical mobile communication system (AeroMACS) as a future communication system. Built on the IEEE 802.16 standard and designed for airport terminals, AeroMACS enables the transmission of high-resolution images

Real-Time Ground-Based Flight Data and Cockpit Voice Recorder: Implementation Scenarios and Feasibility Analysis, First Edition. Mustafa M. Matalgah and Mohammed Ali Alqodah.
© 2024 The Institute of Electrical and Electronics Engineers, Inc. Published 2024 by John Wiley & Sons, Inc.

and videos through a high-speed wireless connection, facilitating extensive information sharing. Furthermore, AeroMACS supports IP operation, reducing the costs associated with system setup and software development. With encryption across multiple protocol layers, AeroMACS also offers enhanced security compared to the standard aeronautical mobile communication system. Given its numerous advantages, AeroMACS holds significant potential as a platform for implementing IoT in the aviation industry [Morioka et al., 2020].

8.1.1 SWIM Definition

SWIM is an integrated framework that encompasses Standards, Infrastructure, and Governance to facilitate the efficient management and communication of aeronautical information within the air navigation system (ANS), as depicted in Figure 8.1. It enables authorized parties to exchange ANS-related information through standardized and interoperable services. The primary technical program for information management, known as GANP ASBU B2-SWIM, is established by ICAO. This program defines SWIM services and infrastructure, as well as establishes data models and protocols based on aviation intranet and internet technologies to ensure interoperability.

The formation of the ICAO Information Management Panel (IMP) in 2015 has been instrumental in the development of effective concepts for implementing SWIM. One of their key responsibilities is to establish standardized information exchange formats within the ANS, ensuring a global and harmonized approach. Another important concept related to SWIM is the notion of the SWIM Region. This refers to a specific geographic area where a collective of States and ATM stakeholders have reached an agreement on common regional governance. This governance framework supports the implementation of SWIM practices. The overarching objective of SWIM is to facilitate the global exchange of aviation

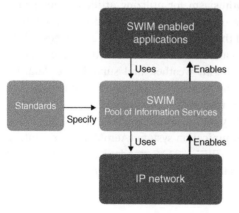

Figure 8.1 SWIM.
Source: EUROCONTROL
[2021]/EUROCONTROL.

information. By delivering relevant and timely content to intended users, SWIM aims to enable collaborative processes in the future. This concerted effort aligns with the goal of enhancing communication and coordination in the aviation industry [EUROCONTROL, 2021, Uniting Aviation, 2019].

8.1.2 SWIM Principles

In a Manual on SWIM Concept, Doc 10039 AN/511, ICAO ICAO [2015] introduced the key SWIM principles, which aim to meet the global requirements of ATM by drawing upon successful strategies from various information communities. The primary objective of SWIM is to establish an interoperable system that enables users to access information that is both valuable and easily understandable by all involved parties. Interoperability, in this context, refers to the seamless sharing of data between different systems and potentially across different organizations. It encompasses the ability to effectively communicate and exchange data, as well as comprehend the exchanged information in a meaningful manner. To support network-centric ATM operations, it is crucial that the information is of sufficient quality, available when needed and delivered to the appropriate location. To achieve these goals, the following SWIM principles should be followed:

1. *Separation of information provision and consumption*: Wherever possible, participants in the ATM network should act as both producers and consumers of information. This separation allows for flexibility and enhances the exchange of information.
2. *Loose system coupling*: Each component of the system should have minimal knowledge or dependency on the definitions of other distinct components. By minimizing barriers between systems and applications, interfaces become compatible, and facilitating seamless integration.
3. *Use of open standards*: Open standards are publicly available and come with various usage rights. They may also have specific properties that describe their design phase, such as an open process. Utilizing open standards ensures transparency, accessibility, and broader adoption.
4. *Use of interoperable services*: By analyzing the processes and requirements of different business domains, the necessary functionality is developed, packaged, and implemented as a suite of interoperable services. These services can be flexibly utilized across multiple separate systems, promoting seamless integration and collaboration.

Following these SWIM principles enables effective implementation, enhancing the exchange of aeronautical information and supporting efficient decision-making within the ATM domain.

8.1.3 SWIM Layers

To define information sharing via SWIM, the Manual on SWIM Concept, Doc 10039 AN/511 [ICAO, 2015] provided a five-layered SWIM global interoperability framework. The framework relies on several overlapping descriptions of SWIM developed by interested parties. The layers describe the functions, the combination of representative standards, and the interoperability mechanisms. The current description of the SWIM global interoperability framework conforms to the solutions being explored for the aerial portion and emphasizes the technical aspects of the ground–ground SWIM segment. The global interoperability framework, as illustrated in Figure 8.2, comprises the following layers:

1. SWIM-enabled applications of information providers and information consumers around the globe. Individuals and organizations, such as air traffic managers and airspace users, will interact using applications that interoperate through SWIM.
2. Information exchange services defined for each ATM information domain and for cross-domain purposes, where opportune, following governance specifications and agreed upon by SWIM stakeholders. SWIM-enabled applications will use information exchange services for interaction.
3. Information exchange models using subject-specific standards for sharing information for the above information exchange services. The information exchange models define the syntax and semantics of the data exchanged by applications.

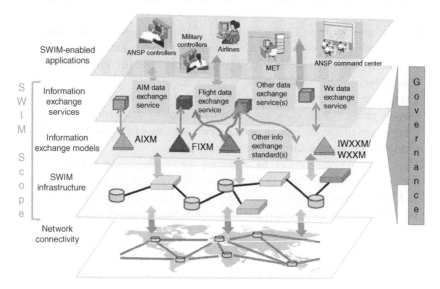

Figure 8.2 SWIM global interoperability framework. Source: ICAO [2015]/International Civil Aviation Organization.

4. SWIM infrastructure for sharing information. It provides the core services such as interface management, request–reply and publish–subscribe messaging, service security, and enterprise service management.
5. Network connectivity provides consolidated telecommunications services, including hardware. This infrastructure is a collection of the interconnected network infrastructures of the different stakeholders. These will be private/public Internet Protocol networks.

The primary focus of SWIM is on the three intermediate layers: information exchange services, information exchange models, and SWIM infrastructure, along with their governance. Currently, SWIM identifies a few essential information exchange services of global significance, such as aeronautical information, meteorological information, surveillance information, and flight information. As SWIM evolves, it is anticipated that additional types of domain services and composite services that span multiple domains will be defined. The key objective for stakeholders involved in implementing SWIM is to reach a consensus on a set of information exchange services, taking into account the varying levels of ATM expertise among member states. Additionally, defining the range of options within each service and establishing information exchange standards are important considerations.

8.2 Air-to-Ground (A2G) Future Communication

Based on the insights provided in Jiang et al. [2017], the field of air-to-ground communications can experience significant advancements by leveraging several emerging technologies. These include millimeter wave (mmWave), mobile edge computing (MEC), ultra-wideband (UWB) technology, long-range (LoRa), and WiFi. mmWave technology harnesses the high-frequency (HF) bands above 6 GHz, enabling it to support data-intensive applications with substantial capacity. By integrating mmWave into air-to-ground communications, it becomes possible to facilitate high-speed and large-scale data transfers between airborne and ground-based systems. MEC is a promising technology that enhances air-to-ground communications by bringing computational capabilities closer to the network edge. This approach reduces latency and improves overall system performance, particularly in time-sensitive applications that require real-time data processing and decision-making. UWB technology offers precise and accurate positioning capabilities, making it invaluable for applications involving location tracking or navigation assistance. By incorporating UWB into air-to-ground communications, situational awareness can be enhanced, leading to more efficient operations. LoRa technology provides low-power, wide-area network (LPWAN) connectivity, enabling LoRa communication with minimal energy consumption. Implementing LoRa in air-to-ground communications extends coverage areas and supports remote sensing and monitoring applications. WiFi, as a widely adopted wireless communication technology, can also contribute to air-to-ground communications. By leveraging existing WiFi infrastructure and protocols, reliable and high-speed connections can be established between airborne and ground-based systems. This facilitates seamless data exchange and enables the deployment of various applications.

8.3 Advancements in Air-to-Air (A2A) Communication for Aviation

Air-to-Air (A2A) communication plays a crucial role in future aviation systems, encompassing tasks such as separation assurance, collision avoidance, and providing passengers with in-flight Internet access for entertainment purposes. In this section, we will delve into the four primary technologies employed in A2A communication, as highlighted in Zhang et al. [2019]: airborne collision avoidance system (ACAS), airborne separation assurance systems (ASAS), L-DACS1 A2A Mode, and free-space optical (FSO) communications.

8.3.1 Airborne Collision Avoidance System (ACAS)

The ACAS facilitates direct, low-latency A2A communication between aircraft, operating independently of ground-based equipment and air traffic control (ATC) systems. It provides pilots with advanced awareness of nearby aircraft that pose a collision risk. ACAS utilizes Secondary Surveillance Radar (SSR) transponder signals to generate resolution advisories (RAs) when it detects an imminent collision, alerting the pilots. ACAS encompasses three types: ACAS-I, which provides traffic advisories (TAs) but no RAs; ACAS-II, which offers TAs and RAs in the vertical direction; and ACAS-III, which provides TAs and RAs in both the vertical and horizontal directions (though ACAS-III is not yet fully deployed). While current ACAS implementations solely facilitate emergency communication between pilots and lack data transfer capabilities for passenger applications, there is potential for future research to explore expanding these systems to enable close data connections between adjacent aircraft.

8.3.2 Airborne Separation Assurance Systems (ASAS)

ASAS furnish flight information on adjacent traffic to aid pilots in maintaining safe separation distances from other aircraft. ASAS aligns with the concept of "Free Flight," empowering pilots to optimize their flight paths, reduce journey durations, and enhance airspace utilization. ASAS permits the exchange of flight information among aircraft at a low data rate. Similar to ACAS, ASAS does not cater to passenger services. To facilitate pilot-controlled self-separation and maneuvering, ASAS employs airborne surveillance and separation assurance processing equipment to analyze surveillance reports from various sources, evaluating the target data based on predetermined criteria.

8.3.3   L-DACS1 A2A Mode

The A2A mode of L-DACS1 focuses on the periodic transmission of surveillance data, while also allowing for a limited number of nonperiodic A2A messages. L-DACS1 utilizes a self-adaptive slotted TDMA protocol and encompasses services such as paired approach, self-separation, and ATC surveillance. The maximum allowable throughput for network users is up to 273 kbps. However, the current implementation of L-DACS1 does not support the transfer of passenger data, but future iterations are expected to incorporate this functionality.

8.3.4   Free-Space Optical (FSO) Communications

FSO communications employ laser diodes as transmitters, enabling high-speed communication, reaching up to 600 Mbps, between aircraft or between aircraft and satellites in Mobile Ad hoc Network (MANET) applications. However, the suitability of FSO for Air-to-Ground (A2G) communications requires further investigation to ensure safety considerations, particularly concerning eye safety. FSO's directional and license-free characteristics make it appealing for aviation communication, overcoming the bandwidth limitations of conventional radio frequency (RF) communication. Nonetheless, FSO communications are sensitive to mobility as they require LOS alignment for reliable transmission. Combining the aircraft's built-in GPS system with low-latency FSO technology can address the challenges of precise pointing and tracking. It's worth noting that FSO links necessitate a LOS channel, which may not be an issue in the stratosphere. Consequently, FSO communication links between aircraft hold immense potential for developing A2A networks for aircraft tracking and collision avoidance. Furthermore, the integration of hybrid FSO/RF links shows promise for future aircraft communications surveillance systems as well.

8.4   Emerging Technologies Shaping Aviation Communication

8.4.1   Single-Pilot Operations (SPOs)

In the realm of future aviation operations, single-pilot operations (SPOs) have garnered attention as a concept explored in Bailey et al. [2017]. SPOs involve a single pilot operating an aircraft with the assistance of onboard automation and a ground operator. In this context, connectivity serves as a vital technology to support safe aircraft operations. Robust and reliable airplane-ground connectivity plays a crucial role in monitoring and facilitating remote aircraft operations. SPOs encompass various research aspects, including massive data transfer, where numerous sensors on aircraft are employed to monitor and predict maintenance requirements. The utilization of sensors aids in reducing maintenance costs and extending the lifespan of aircraft.

8.4.2 Troposcatter Communications

Troposcatter communications, which have been employed for long-distance communications in remote areas, have been extensively used by military organizations and others prior to the advent of satellite communications. Troposcatter leverages the scattering of transmitted signal energy by tropospheric inhomogeneities for communication. These systems typically utilize high-gain antennas with narrow beam widths and employ considerable transmission power. Recent interest in troposcatter research has focused on developing more precise channel models and transmission techniques such as multiple-input multiple-output (MIMO) to enhance reliability. Troposcatter enables practical Mbps data rates while maintaining a reasonable level of security [Erturk et al., 2019].

8.4.3 Near Vertical Incidence Skywave (NVIS) Communications

Near vertical incidence skywave (NVIS) Communications employ a portion of the high-frequency (HF) band, typically ranging from 0.5 to 10 MHz. NVIS utilizes the technique of sending waves at steep elevation angles during the day and relying on ionospheric refraction to reach a radius of several hundred kilometers from the transmitter. This technique, known as "near-vertical incidence," has gained attention recently. NVIS communications are well-suited for local aviation applications and can even be implemented in certain drone applications [Erturk et al., 2019].

8.5 Machine Learning in Future Communications

The potential of machine learning in wireless communication systems, particularly in aerospace communication systems, was explored in Erturk et al. [2019]. Machine learning applications have predominantly revolved around cognitive radio networks (CRNs), which intelligently adapt to identify free channels within a specific frequency spectrum range to accommodate a larger number of users sharing the wireless bandwidth. Machine learning algorithms can optimize spectrum efficiency, power allocation, antenna selection, and beamforming in CRNs. Moreover, machine learning techniques find relevance in navigation, surveillance, massive MIMO, femto/small cells, heterogeneous networks, smart grid, energy harvesting, and device-to-device communications, all of which are part of the evolving landscape of 5G networks. The integration of machine learning algorithms can enhance the reliability, efficiency, and resource utilization of aircraft communication systems, benefiting navigation, surveillance, and other related applications.

8.6 Summary

In this chapter, the focus is on advancements in aviation communication that offer potential improvements for FDR and CVR data streaming. The chapter begins by introducing the SWIM platform developed by the ICAO for the sharing of aeronautical data. SWIM enables standardized and interoperable services, facilitating the seamless integration of aeronautical information systems.

The chapter highlights the challenges faced in integrating ATM information systems due to diverse data formats and exchange protocols. To overcome these challenges, ICAO has standardized the SWIM platform, which ensures the sharing of aeronautical big data, including flight, weather, and surveillance data, through standardized formats and protocols. The AeroMACS is introduced as a future communication system built on the IEEE 802.16 standard. AeroMACS enables high-speed wireless transmission of high-resolution images and videos, supporting extensive information sharing in the aviation industry. It also offers enhanced security and supports IP operation, reducing setup costs.

The chapter then explores advancements in A2G communication, highlighting emerging technologies such as mmWave, MEC, UWB technology, LoRa, and WiFi. These technologies can enhance data transfers, reduce latency, improve situational awareness, extend coverage areas, and establish reliable connections between airborne and ground-based systems.

Advancements in A2A communication for aviation are discussed, focusing on four primary technologies: ACAS, ASAS, L-DACS1 A2A Mode, and FSO communications. ACAS provides collision risk awareness and resolution advisories to pilots, while ASAS enables pilots to optimize flight paths and maintain safe separation distances. L-DACS1 A2A Mode allows for surveillance data transmission, and FSO communications offer high-speed communication between aircraft or aircraft and satellites.

The chapter concludes with a discussion of emerging technologies shaping aviation communication, including SPOs that rely on robust airplane-ground connectivity for remote operations. Troposcatter communications, which utilize tropospheric scattering for long-distance communication, and NVIS communications, which use skywave propagation for communication, are also highlighted as promising technologies.

References

Randall E. Bailey, Lynda J. Kramer, Kellie D. Kennedy, Chad L. Stephens, and Timothy J. Etherington. An assessment of reduced crew and single pilot operations in commercial transport aircraft operations. In *2017 IEEE/AIAA 36th Digital Avionics Systems Conference (DASC)*, pages 1–15, 2017. doi: 10.1109/DASC. 2017.8101988.

M. Cenk Erturk, Hosseinali Jamal, and David W. Matolak. Potential future aviation communication technologies. In *2019 IEEE/AIAA 38th Digital Avionics Systems Conference (DASC)*, pages 1–10, 2019. doi: 10.1109/DASC43569.2019.9081679.

EUROCONTROL. System Wide Information Management (SWIM) factsheet, 2021.

ICAO. Manual on System Wide Information Management (SWIM) concept, 2015. ICAO Doc 10039.

Chunxiao Jiang, Haijun Zhang, Yong Ren, Zhu Han, Kwang-Cheng Chen, and Lajos Hanzo. Machine learning paradigms for next-generation wireless networks. *IEEE Wireless Communications*, 24(2):98–105, 2017. doi: 10.1109/MWC.2016. 1500356WC.

Kazuyuki Morioka, Xiaodong Lu, Junichi Naganawa, Norihiko Miyazaki, Naruto Yonemoto, Yasuto Sumiya, and Akiko Kohmura. Service assurance packet-scheduling algorithm for a future aeronautical mobile communication system. *Simulation Modelling Practice and Theory*, 102:102059, 2020. ISSN 1569-190X. doi: 10.1016/j.simpat.2019.102059. URL https://www.sciencedirect .com/science/article/pii/S1569190X1930190X. Special Issue on IoT, Cloud, Big Data and AI in Interdisciplinary Domains.

Uniting Aviation. System wide information management progression. In *The APAC Region*, 2019. URL https://unitingaviation.com/regions/system-wide-information-management-progression-in-the-apac-regions/.

Jiankang Zhang, Taihai Chen, Shida Zhong, Jingjing Wang, Wenbo Zhang, Xin Zuo, Robert G. Maunder, and Lajos Hanzo. Aeronautical ad hoc networking for the internet-above-the-clouds. *Proceedings of the IEEE*, 107(5):868–911, 2019. doi: 10.1109/JPROC.2019.2909694.

Appendix A

A.1 Useful MATLAB Codes

A.1.1 Iridium Satellite Constellation Viewer

```
%Iridium Satellite Constellation Viewer
close all;
mission.StartDate = datetime(2023,5,1,12,0,0);
%Simulation duration
mission.Duration = hours(1);
mission.scenario = satelliteScenario(mission.StartDate,...
    mission.StartDate + mission.Duration, 60);
mission.viewer = satelliteScenarioViewer(mission.scenario);
constellation.obj = walkerStar(mission.scenario, 781e3+6378.14e3,...
    86.4, 66, 6, 2, Name="Iridium");
%Play the scenario.
play(mission.scenario);
```

A.1.2 Iridium Satellite Constellation Footprints

```
%Illustration of RF overlapping footprints of the
%Iridium NEXT satellite constellation
close all;
mission.StartDate = datetime(2023,5,1,12,0,0);
mission.Duration = hours(1);

mission.scenario = satelliteScenario(mission.StartDate, ...
    mission.StartDate + mission.Duration, 60);
mission.viewer = satelliteScenarioViewer(mission.scenario);
constellation.obj = walkerStar(mission.scenario, 781e3+6378.14e3,...
    86.4, 66, 6, 2, Name="Iridium");
```

Real-Time Ground-Based Flight Data and Cockpit Voice Recorder: Implementation Scenarios and Feasibility Analysis, First Edition. Mustafa M. Matalgah and Mohammed Ali Alqodah.
© 2024 The Institute of Electrical and Electronics Engineers, Inc. Published 2024 by John Wiley & Sons, Inc.

```
sensor.HalfAngle = 61.5; %deg
sensor.Names = constellation.obj.Name + " satellite";
sensor.obj = conicalSensor(constellation.obj,...
    MaxViewAngle=sensor.HalfAngle*2, MountingLocation=[0 0 10],...
    Name=sensor.Names);
sensor.FOV.obj = fieldOfView(sensor.obj);

mission.viewer.PlaybackSpeedMultiplier = 200;
play(mission.scenario);
```

A.1.3   Large Satellite Constellation Implementation for Ground-Based FDR/CVR Recorders

```
%large Low Earth Orbit (LEO) satellite constellation to
%transmit Flight Data Recorder (FDR) and Cockpit Voice Recorder
    %(CVR) data for
%an aircraft flying from New York to Chicago.
%Using (Satellite Communications Toolbox) Examples.


%Create Satellite Scenario
startTime = datetime(2023,5,4,18,27,57);
stopTime = startTime + hours(2.5);
sampleTime = 60;                          %Seconds
sc = satelliteScenario(startTime,stopTime,sampleTime,...
    "AutoSimulate",false)
%Add Large Constellation of Satellites

sat = satellite(sc,"largeConstellation.tle");

numSatellites = numel(sat)

%Airports
%Create the airports properties in the scenario.

airportName = ["JFK International (New York)";...
    "O'Hare International Airport (Chicago)"];
airportLat = [40.6413;41.9675];
airportLon = [-73.7781;-87.8930];
%Add the airports to the scenario using the groundStation
    %(Satellite Communications Toolbox) function.
```

```
airports = groundStation(sc,airportLat,airportLon,Name=airportName);
%Aircraft
%Create a geoTrajectory

startLLA = [airportLat(1) airportLon(1) 10600];
endLLA = [airportLat(2) airportLon(2) 10600];
timeOfTravel = [0 seconds(sc.StopTime-sc.StartTime)];
sampleRate   = 1/30;

trajectory = geoTrajectory([startLLA;endLLA],timeOfTravel,...
    SampleRate=sampleRate,...
    AutoPitch=true,AutoBank=true);

[positionLLA,orientation] = trajectory();

[aircraftPosition,aircraftOrientation] = lookupPose(trajectory,...
    trajectory.TimeOfArrival(1):(1/trajectory.SampleRate):...
      trajectory.TimeOfArrival(end));
%Generate a timetable for the aircraft position and orientation.
  %Use the retime function to interpolate
%the position and orientation into the same time step as the
  %scenario, every 60 seconds.

aircraftPositionTT = timetable(aircraftPosition,...
    StartTime=sc.StartTime,...
    TimeStep=seconds(1/trajectory.SampleRate),...
    VariableNames="Lat-Lon-Alt");
aircraftOrientationTT = timetable(compact(aircraftOrientation),...
    StartTime=sc.StartTime,...
    TimeStep=seconds(1/trajectory.SampleRate),...
    VariableNames="Orientation");

aircraftPositionTT = retime(aircraftPositionTT,'regular',...
'linear',...
    TimeStep=seconds(sc.SampleTime));
aircraftOrientationTT = retime(aircraftOrientationTT,'regular',...
    'nearest',TimeStep=seconds(sc.SampleTime));

aircraft.obj = satellite(sc,aircraftPositionTT,...
    CoordinateFrame="geographic", Name="Aircraft");
aircraft.obj.MarkerColor = "green";
```

```
aircraft.obj.Orbit.LineColor = "green";
gsSource=aircraft.obj;

%determine the  target ground station

gsTarget = groundStation(sc,17.4351,78.3824, ...
    "Name","Target Ground Station");

%Determine Elevation Angles of Satellites with Respect to Ground
    %Stations
%Calculate the scenario state corresponding to StartTime.
advance(sc);

%Retrieve the elevation angle of each satellite with respect to
    %the ground
%stations.
[~,elSourceToSat] = aer(gsSource,sat);
[~,elTargetToSat] = aer(gsTarget,sat);

%Determine the elevation angles that are greater than or equal
    %to 30
%degrees.
elSourceToSatGreaterThanOrEqual30 = (elSourceToSat >= 30)';
elTargetToSatGreaterThanOrEqual30 = (elTargetToSat >= 30)';

%Determine Best Satellite for Initial Access to Constellation

trueID = find(elSourceToSatGreaterThanOrEqual30 == true);

[~,~,r] = aer(sat(trueID), gsTarget);

%Determine the index of the element in r bearing the minimum
    %value.
[~,minRangeID] = min(r);

%Determine the element in trueID at the index minRangeID.
id = trueID(minRangeID);
nodes = {gsSource sat(id)};

%Determine Remaining Nodes in Path to Target Ground Station
earthRadius = 6378137;
    %meters
```

```matlab
altitude = 500000;
   %meters
horizonElevationAngle = asind(earthRadius/(earthRadius + ...
altitude))...
   - 90 %degrees

%Minimum elevation angle of satellite nodes with respect to the
   %prior
%node.
minSatElevation = -15; %degrees

%Flag to specify if the complete multi-hop path has been found.
pathFound = false;

%Determine nodes of the path in a loop. Exit the loop once the
   %complete
%multi-hop path has been found.
while ~pathFound
    %Index of the satellite in sat corresponding to current node
        %is
    %updated to the value calculated as index for the next node in
        %the
    %prior loop iteration. Essentially, the satellite in the next
        %node in
    %prior iteration becomes the satellite in the current node
        %in this
    %iteration.
    idCurrent = id;

    %This is the index of the element in
        %elTargetToSatGreaterThanOrEqual30
    %tells if the elevation angle of this satellite is at least
        %30 degrees
    %with respect to "Target Ground Station". If this element is
        %true, the
    %routing is complete, and the next node is the target ground
        %station.
    if elTargetToSatGreaterThanOrEqual30(idCurrent)
        nodes = {nodes{:} gsTarget}; %#ok<CCAT>
        pathFound = true;
        continue
    end
```

```
%If the element is false, the path is not complete yet. The
    %node
%in the path must be determined from the constellation.
    %Determine
%which satellites have elevation angle that is greater than
    %or equal
%to -15 degrees with respect to the current node. To do this,
    %first
%determine the elevation angle of each satellite with respect
    %to the
%current node.
[~,els] = aer(sat(idCurrent),sat);

%Overwrite the elevation angle of the satellite with respect
    %to itself
%to be -90 degrees to ensure it does not get re-selected as
    %the next
%node.
els(idCurrent) = -90;

%Determine the elevation angles that are greater than or equal
    %to -15
%degrees.
s = els >= minSatElevation;

%Find the indices of the elements in s whose value is true.
trueID = find(s == true);

%These indices are essentially the indices of satellites in
    %sat whose
%elevation angle with respect to the current node is greater
    %than or
%equal to -15 degrees. Determine the range of these satellites
    %to
%"Target Ground Station".
[~,~,r] = aer(sat(trueID), gsTarget);

%Determine the index of the element in r bearing the minimum
    %value.
[~,minRangeID] = min(r);
```

```
%Determine the element in trueID at the index minRangeID.
id = trueID(minRangeID);


%This is the index of the best satellite among those in sat to
    %be used
%for the next node in the path. Append this satellite to the
    %'nodes'
%variable.
    nodes = {nodes{:} sat(id)}; %#ok<CCAT>
end


%Determine Intervals When Calculated Path Can Be Used
sc.AutoSimulate = true;


%Add an access analysis with the calculated nodes in the path.
%Set LineColor of the access visualization to red.
ac = access(nodes{:});
ac.LineColor = "red";

%Determine the access intervals using the accessIntervals function
intvls = accessIntervals(ac)

%Visualize Path
pointAt(sat,aircraft.obj);

v = satelliteScenarioViewer(sc,"ShowDetails",false);
sat.MarkerSize = 6; %Pixels
campos(v,40,-75);        %Latitude and longitude in degrees

%Play the scenario.
play(sc);
```

Note: Upon copying the Matlab code straight from the published book PDF file and pasting it into Matlab m-file, when executing it there will be an issue with the identification of the "~" symbol as used in the published book's appendix. Matlab returns an error message stating "invalid text character." However, after replacing the copied "~" symbol with the standard tilde "~" from the keyboard, the code will be executed correctly.
This issue will not arise in when the user types the code directly from the computer keyboard. However, it could potentially cause problems only for those attempting to copy and paste the code from the e-book.

Index

*Real-Time Ground-Based Flight Data and Cockpit Voice Recorder: Implementation Scenarios and
Feasibility Analysis*, First Edition. Mustafa M. Matalgah and Mohammed Ali Alqodah.
© 2024 The Institute of Electrical and Electronics Engineers, Inc. Published 2024 by John Wiley & Sons, Inc.

Printed and bound by CPI Group (UK) Ltd, Croydon, CR0 4YY

16/04/2025

14658601-0001